SEQUENTIAL ANALYSIS

WILEY MATHEMATICAL STATISTICS SERIES

Walter A. Shewhart, Editor

Mathematical Statistics
  HOEL—Introduction to Mathematical Statistics.
  WALD—Sequential Analysis.

Applied Statistics
  DODGE and ROMIG—Sampling Inspection Tables.
  RICE—Control Charts.

Related Books of Interest to Statisticians
  HAUSER and LEONARD—Government Statistics for Business Use.

SEQUENTIAL ANALYSIS

By

ABRAHAM WALD

Professor of Mathematical Statistics
Columbia University

NEW YORK
JOHN WILEY & SONS, INC.
CHAPMAN & HALL, LTD.
LONDON

PREFACE

This book presents the theory of a recently developed method of statistical inference, that of sequential analysis. An effort has been made to keep the exposition on a level that will make most of the book, with the exception of the Appendix, understandable to readers whose mathematical background does not go beyond college algebra and a first course in calculus. Some knowledge of probability and statistics is desirable for the understanding of the book, although not essential, for a brief review is given of the fundamental concepts, such as random variables, probability distributions, and statistical hypotheses.

To facilitate the reading of the book for those who have no advanced mathematical training, some concessions are made to generality and occasionally even to rigor. Furthermore, mathematical derivations of somewhat intricate nature are put into the Appendix, the reading of which may be omitted without impairing the understanding of the rest of the book.

This book contains an expanded exposition of the ideas and results I published in two technical papers on this subject, one of which appeared in 1944 and the other in 1945, as well as some further developments. Such developments, for example, are: the discussion of multi-valued decisions and estimation in Part III; improvements in the limits for the average number of observations required by a sequential test; and limits for the effect of grouping in the binomial case. Some recent results of M. A. Girshick are included and, in the discussion of certain applications in Part II, use is made of some simplifications contained in a publication of the Statistical Research Group of Columbia University dealing with these applications.

Nearly all tables in the book were computed by the Statistical Research Group of Columbia University while I was a consultant to the group. A few sections of my two forementioned publications have been incorporated in this book, mostly in the Appendix, without substantial changes.

I wish to express my indebtedness to Milton Friedman and W. Allen Wallis, who proposed the problem of sequential analysis to me in March, 1943. It was their clear formulation of the problem that gave me the incentive to start the investigations leading to the present

developments.   I also wish to express my thanks to the Social Science
Research Council for their help, which facilitated the publication of
this book.   I am indebted to Mr. Mortimer Spiegelman of the Metro-
politan Life Insurance Company for his careful reading of the manuscript and for making several valuable suggestions.   Thanks are due
also to Mrs. E. Bowker who prepared the manuscript with particular
care.

A. W.

Columbia University
March, 1947

CONTENTS

INTRODUCTION

PART I. GENERAL THEORY

Chapter 1. ELEMENTS OF THE CURRENT THEORY OF TESTING STATISTICAL HYPOTHESES

Chapter 2. SEQUENTIAL TEST OF A STATISTICAL HYPOTHESIS: GENERAL DISCUSSION

Chapter 3. THE SEQUENTIAL PROBABILITY RATIO TEST FOR TESTING A SIMPLE HYPOTHESIS H_0 AGAINST A SINGLE ALTERNATIVE H_1

Chapter 4. OUTLINE OF A THEORY OF SEQUENTIAL TESTS OF SIMPLE AND COMPOSITE HYPOTHESES AGAINST A SET OF ALTERNATIVES

CONTENTS

INTRODUCTION

Sequential analysis is a method of statistical inference whose characteristic feature is that the number of observations required by the procedure is not determined in advance of the experiment. The decision to terminate the experiment depends, at each stage, on the results of the observations previously made. A merit of the sequential method, as applied to testing statistical hypotheses, is that test procedures can be constructed which require, on the average, a substantially smaller number of observations than equally reliable test procedures based on a predetermined number of observations.

This book presents the theory of a particular method of sequential analysis, the so-called sequential probability ratio test, which was devised by the author in 1943 mainly for the purpose of testing statistical hypotheses. A comparison of this particular sequential test procedure with any other (sequential or non-sequential) is shown, in Section A.7, to effect the greatest possible saving in the average number of observations, when used for testing a simple hypothesis against a single alternative. The sequential probability ratio test frequently results in a saving of about 50 per cent in the number of observations over the most efficient test procedure based on a fixed number of observations.

The first idea of a sequential test procedure, i.e., a test for which the number of observations is not determined in advance but is dependent on the outcome of the observations as they are made, goes back to H. F. Dodge and H. G. Romig [1] who constructed a double sampling procedure. According to this scheme the decision whether or not a second sample should be drawn depends on the outcome of the observations in the first sample. Whereas this method allows for only two samples, Walter Bartky devised a multiple sampling scheme for the particular case of testing the mean of a binomial distribution.[2] His scheme is closely related to the test procedure that results from the application of the sequential probability ratio test to this particular case. The reason that Dodge and Romig introduced their double

[1] H. F. Dodge and H. G. Romig, "A Method of Sampling Inspection," *The Bell System Technical Journal*, Vol. 8 (1929), pp. 613–631.
[2] Walter Bartky, "Multiple Sampling with Constant Probability," *The Annals of Mathematical Statistics*, Vol. 14 (1943), pp. 363–377.

1

sampling method, and Bartky his multiple sampling scheme was, of course, the recognition of the fact that they require, on the average, a smaller number of observations than "single" sampling.

The occasional practice of designing a large scale experiment in successive stages may be regarded as a forerunner of sequential analysis. The idea of such chain experiments was briefly discussed by Harold Hotelling.[3] A very interesting example of this type is the series of sample censuses of area of jute in Bengal carried out under the direction of P. C. Mahalanobis.[4] Sample censuses, steadily increasing in size, were taken primarily for the purpose of obtaining preliminary information about the parameters to be estimated. This information was then used for designing the final sampling of the whole immense jute area in Bengal.

The problem of sequential analysis arose in the Statistical Research Group of Columbia University [5] in connection with some comments made by Captain G. L. Schuyler of the Bureau of Ordnance, Navy Department. Milton Friedman and W. Allen Wallis recognized the great potentialities and the far-reaching consequences that sequential analysis might have for the further development of theoretical statistics. In particular, they conjectured that a sequential test procedure might be constructed which would control the possible errors committed by wrong decisions exactly to the same extent as the best current procedure based on a predetermined number of observations, and at the same time would require, on the average, a substantially smaller number of observations than the fixed number of observations needed for the current procedure.[6] Friedman and Wallis also exhibited a few examples of sequential modifications of current test procedures resulting, in some cases, in an increase of efficiency. It was at this stage that they proposed the problem of sequential analysis to the author. This gave the incentive for the author's investigations which then led to the development of the sequential probability ratio test.

[3] Harold Hotelling, "Experimental Determination of the Maximum of a Function," *The Annals of Mathematical Statistics*, Vol. 12 (1941), pp. 20–45.

[4] P. C. Mahalanobis, "A Sample Survey of the Acreage under Jute in Bengal, with Discussion on Planning of Experiments," *Proceedings of the 2nd Indian Statistical Conference*, Calcutta, Statistical Publishing Society (1940).

[5] During World War II the Statistical Research Group operated under a contract with the Office of Scientific Research and Development and was directed by the Applied Mathematics Panel of the National Defense Research Committee.

[6] Bartky's multiple sampling scheme for testing the mean of a binomial distribution provides an example of such a sequential test. His results were not known to Friedman and Wallis at that time, since they were published nearly a year later.

Because of the usefulness of the sequential probability ratio test in development work on military and naval equipment, it was classified Restricted within the meaning of the Espionage Act. The author was requested to submit his findings in a restricted report [7] dated September, 1943.[8] In this report the sequential probability ratio test is devised and the basic theory is given. To facilitate the use of this new technique by the Army and the Navy, the Statistical Research Group issued a second report in July, 1944, which gives an elementary non-mathematical exposition of the applications of the sequential probability ratio test and contains a considerable number of tables, charts, and computational simplifications to facilitate applications.[9]

Further advances in the theory of the sequential probability ratio test were made in 1944. The operating characteristic (OC) curve of the sequential probability ratio test for the case of a binomial distribution was found by Milton Friedman and George W. Brown (independently of each other), and slightly earlier by C. M. Stockman in England.[10] The author then obtained the general OC curve for any sequential probability ratio test.[11] A few months later a general theory of cumulative sums was developed [12] which gives not only the OC curve of any sequential probability ratio test but also the characteristic function of the number of observations required by the test and various other results.

The material in the author's report together with the new advances made in 1944 were published by him in a paper, "Sequential Tests of Statistical Hypotheses," in *The Annals of Mathematical Statistics*, June, 1945. The Statistical Research Group issued a revised edition [13] of its

[7] Abraham Wald, "Sequential Analysis of Statistical Data: Theory," a report submitted by the Statistical Research Group, Columbia University, to the Applied Mathematics Panel, National Defense Research Committee, Sept., 1943.

[8] The restricted classification was removed in May, 1945.

[9] Harold Freeman, "Sequential Analysis of Statistical Data: Applications," a report submitted by the Statistical Research Group, Columbia University, to the Applied Mathematics Panel, National Defense Research Committee, July, 1944.

[10] C. M. Stockman, "A Method of Obtaining an Approximation for the Operating Characteristic of a Wald Sequential Probability Ratio Test Applied to a Binomial Distribution," (British) Ministry of Supply, Advisory Service on Statistical Method and Quality Control, Technical Report, Series "R," No. Q.C./R/19.

[11] Abraham Wald, "A General Method of Deriving the Operating Characteristics of any Sequential Probability Ratio Test," unpublished memorandum submitted to the Statistical Research Group, Columbia University, April, 1944.

[12] Abraham Wald, "On Cumulative Sums of Random Variables," *The Annals of Mathematical Statistics*, Vol. 15 (Sept., 1944).

[13] The authorship of the revised edition, which was published by the Columbia University Press, Sept., 1945, is ascribed to the group as a whole.

original report. The revised edition includes a discussion of the operating characteristic and average sample number curves for various
applications of the sequential probability ratio test.

Independently of the development in this country and about the
same time, G. A. Barnard recognized the merits of a sequential method
of testing.[14] He treated the problem of double dichotomies, using a
sequential method of testing which, however, differs from the one that
results from the application of the sequential probability ratio test.

This book consists of three parts and an Appendix. Part I contains
a discussion of the general theory of the sequential probability ratio
test. Part II discusses applications of the general theory given in
Part I. These applications are given primarily to illustrate the general theory and to bring out some points of theoretical interest which
are specific to these applications. Accordingly, computational simplifications are not stressed much and hardly any tables are given.[15]
Part III outlines briefly a possible approach to the problem of sequential multi-valued decisions and estimation. This field is largely un-
explored and further progress is still a matter of future developments.
To facilitate the use of the book by readers with no advanced mathematical training, mathematical derivations of somewhat intricate nature are included in the Appendix.

[14] G. A. Barnard, "Economy in Sampling with Reference to Engineering Experimentation," (British) Ministry of Supply, Advisory Service on Statistical Method
and Quality Control, Technical Report, Series "R," No. Q.C./R/7.

[15] For a more complete and detailed discussion of these applications the reader
is referred to the revised edition of the publication of the Statistical Research
Group mentioned before.

PART I.   GENERAL THEORY

Chapter 1.   ELEMENTS OF THE CURRENT THEORY OF TESTING STATISTICAL HYPOTHESES

1.1   Random Variables and Probability Distributions

1.1.1   Notion of a Random Variable

The outcome of an experiment or the reading of a measurement is usually a variable quantity or, more briefly, a variable, since generally it can take different values. For example, repeated measurements on the length of a bar will yield, in general, different values. Frequently, it will be possible to make probability statements concerning the outcome of an experiment or the reading of a measurement. Consider, for example, the experiment consisting of the throw of a die whose sides are numbered from 1 to 6. Here the outcome of the experiment may be any integral number from 1 to 6. Various probability statements regarding the outcome of the experiment can be made. For example, the probability that the outcome will be equal to 5 is equal to $\frac{1}{6}$, or the probability that the outcome will be less than 4 is equal to $\frac{1}{2}$, and so forth. Probability statements can also be made about the outcome of the following experiment: Suppose that an individual is selected at random from a group of 1000 individuals and that his height is then measured. The probability that the height of the selected individual will be less than 68 inches is equal to $\frac{1}{1000}$ times the number of individuals in the group whose heights are less than 68 inches.

A variable x is called a random variable if for any given value c a definite probability can be ascribed to the event that x will take a value less than c. A general class of experiments where the outcome is a random variable in the sense of the above definition may be described as follows. Consider a class of N objects (or individuals) and some measurable characteristic of these objects, such as weight, diameter, or hardness. Suppose that the value x of this characteristic varies from object to object in the class. The experiment consists in selecting at random one object from the class of N objects, and then measuring the value x of the characteristic of the selected object. Random selection is selection of an object in such a way that each object in the class of N objects has an equal chance of being chosen. The outcome

5

x of such an experiment is a random variable, since a probability can be ascribed to the event that x will take a value less than c, for any given value c. This probability is, in fact, equal to N_c/N, where N_c is the number of objects in the class for which the characteristic under consideration has a value less than c. An interesting special case is that in which the characteristic under consideration can take only two values. Such a situation arises, for instance, in the case of a manufactured product where each unit is classified in one of two categories: defective or non-defective. We shall ascribe the value 0 to a non-defective unit and the value 1 to a defective unit. Then the characteristic under consideration, i.e., the characteristic of being defective or non-defective, can take only the values 0 and 1. Consider a lot consisting of N units and let N_d be the number of defectives in the lot. If the experiment consists in inspecting a single unit drawn at random from the lot, the outcome x of the experiment is a random variable which can take only the values 0 and 1. The probability that $x = 0$ is equal to $(N - N_d)/N$, and the probability that $x = 1$ is equal to N_d/N.

1.1.2 Cumulative Distribution Function (c.d.f.) of a Random Variable

Let x be a random variable and denote by $F(t)$ the probability that x will take a value less than a given value t. Then $F(t)$ is a function of t which is called the cumulative distribution function of x. Since

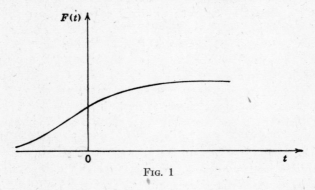

Fig. 1

any probability must lie between 0 and 1, we must have $0 \leqq F(t) \leqq 1$ for all values of t. If t_1 and t_2 are two values such that $t_1 < t_2$, then the probability that $x < t_2$ is greater than or equal to the probability that $x < t_1$, i.e., $F(t_2) \geqq F(t_1)$. In other words, $F(t)$ cannot decrease as t increases. A typical form of a c.d.f. $F(t)$ is shown in Fig. 1 where t is measured along the horizontal axis and $F(t)$ along the vertical axis.

For any given values a and b ($a < b$) we can easily derive the value of the probability that $a \leq x < b$ from the c.d.f. $F(t)$. In fact, the event that $x < a$ and the event that $a \leq x < b$ are mutually exclusive. Hence, the probability that one of these events will occur is equal to the sum of the two probabilities: the probability that $x < a$ and the probability that $a \leq x < b$. Thus, we have

(1:1)   (probability that either $x < a$ or $a \leq x < b$)

$\qquad$ = (probability that $x < a$) + (probability that $a \leq x < b$)

Since the probability that either $x < a$ or $a \leq x < b$ is the same as the probability that $x < b$, we obtain, from (1:1),

(1:2) $\qquad F(b) = F(a) +$ (probability that $a \leq x < b$)

Hence, the probability that $a \leq x < b$ is equal to $F(b) - F(a)$.

A simple interpretation of the c.d.f. $F(t)$ can be given if the random variable x is the value of a measurement on an object selected at random from a given group of N objects. As mentioned in Section 1.1.1, in this case the probability that the observed value of x satisfies some equality or inequality relationship, such as $x = c$, or $x < c$, or $a < x < b$, is equal to the proportion of objects in the group of N objects for which the value of x satisfies the equality or inequality in question. Thus, $F(t)$ is simply equal to the proportion of objects in the group for which $x < t$. With this interpretation of probability, the validity of (1:2) becomes self-evident. It merely says this: The proportion of objects for which $x < b$ is equal to the proportion of objects for which $x < a$ plus the proportion of objects for which $a \leq x < b$. The group of N objects is frequently called *population* or *universe*. So far we have considered only populations which contain a finite number of objects. Such populations are called finite populations.

The interpretation of the probability that a certain relation (equality or inequality) holds as the proportion of objects in the population for which the value of x satisfies that relation proves useful in many instances and we shall employ it frequently. However, if we restrict ourselves to finite populations, such an interpretation is not always possible. In fact, the c.d.f.'s which arise from finite populations are of a special nature. Suppose that N is the number of objects in the population. Then the random variable x can take at most N different values. Let $a_1, \cdots, a_M$ be the different values x can take, arranged in ascending order of magnitude, i.e., $a_1 < a_2 < \cdots < a_M$. Clearly, $M \leq N$. If the value of x is the same for several objects, then $M < N$.

The c.d.f. of x will be a step function of the type shown in Fig. 2. The distribution function makes exactly M jumps and the magnitude of each jump is equal to $1/N$ or an integral multiple of $1/N$. A c.d.f. represented by a continuous curve, as shown in Fig. 1, is certainly not of this type. Thus, if the c.d.f. is given by a continuous curve, the interpretation of probabilities as proportions of a finite population is not possible. However, any c.d.f. can be approximated arbitrarily closely by a c.d.f. arising from a finite population, if the number N of objects in the population is sufficiently large. Thus, any c.d.f. can be

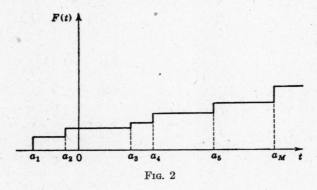

FIG. 2

regarded as a limiting form of a c.d.f. arising from a finite population when the number of objects in the population is increased indefinitely. This means that if we admit infinite populations [1] (populations with infinitely many objects), the interpretation of any probability as a certain proportion of an underlying population is always possible. Of course, the notion of an infinite population is only an abstraction constructed merely for the purpose of simplifying the theory. To give an example of an underlying infinite population, consider a measurement on the length of a bar, the outcome of which is regarded as a random variable x having a c.d.f. $F(t)$. Then the underlying infinite population may be thought of as an infinite sequence of repeated measurements on the length of the bar, and the actually observed measurement is considered an element drawn from this population. Sometimes the underlying population is finite, but the number N of objects in the

---

[1] By an infinite population we mean an ordered infinite sequence of objects, $O_1, O_2, \cdots$, ad inf. A certain measurable characteristic of these objects is considered and the value x of this characteristic is assumed to vary from object to object. By the proportion of objects in the infinite population for which x satisfies a given relation (equality or inequality) we mean the limiting value of the corresponding proportion in the finite population $(O_1, \cdots, O_N)$ as N increases indefinitely.

population is so large that we may find it more convenient to treat the problem as if N were infinity, i.e., as if the population were infinite. Suppose, for example, that we are interested in the height distribution of all male individuals of age 20 and above living in the United S ates. The number of such individuals is so large that considerable mathematical simplification may be achieved by treating the population of such individuals as if it were infinite.

1.1.3 Probability Density Function

Let $F(t)$ be the c.d.f. of a random variable x. As we have seen in Section 1.1.2, the probability that $t - \dfrac{\Delta}{2} \leq x < t + \dfrac{\Delta}{2}$ $(\Delta > 0)$ is given by $F\left(t + \dfrac{\Delta}{2}\right) - F\left(t - \dfrac{\Delta}{2}\right)$. The limiting value $f(t)$ of the ratio

$$\frac{F\left(t + \dfrac{\Delta}{2}\right) - F\left(t - \dfrac{\Delta}{2}\right)}{\Delta}$$

as Δ approaches 0, provided that such a limiting value exists,[2] is called the probability density of the random variable x at the value $x = t$. The probability density $f(t)$ is a function of t and is called the probability density function of the random variable x. It follows from the definition of the probability density $f(t)$ that for small positive values Δ the product $f(t)\Delta$ is a good approximation to the probability that x will lie in the interval $t \pm \dfrac{\Delta}{2}$. A probability density function does not always exist. If the random variable x is discrete, i.e., if x can take only discrete values, the c.d.f. is a step function and no probability density function exists.

The probability that x will take a value within the interval from t_1 to t_2 $(t_1 < t_2)$ can be obtained by integrating the probability density function $f(t)$ from t_1 to t_2; i.e., the probability in question is given by

$$\int_{t_1}^{t_2} f(t)\, dt$$

[2] The existence of the limiting value of $\dfrac{F(t + \Delta) - F(t)}{\Delta}$ is required, where Δ may be positive or negative and may approach 0 in any arbitrary manner. The existence of this limiting value implies the existence of the limiting value of $F\left(t + \dfrac{\Delta}{2}\right) - F\left(t - \dfrac{\Delta}{2}\right)$.

One of the most important probability density functions is the so-called normal probability density function, which is given by

$$(1\!:\!3) \qquad f(t) = \frac{1}{\sqrt{2\pi}\sigma}\, e^{-\frac{1}{2\sigma^2}(t-\mu)^2}$$

where μ and σ are some constant values. If a random variable x has a probability density function $f(t)$ given by (1:3), we say that x is normally distributed, or x has a normal distribution. The shape of a normal curve is shown in Fig. 3, where t is measured along the horizontal axis and $f(t)$ along the vertical axis.

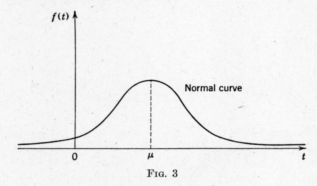

FIG. 3

1.1.4   Discrete Random Variables

A random variable x is called discrete if it can take only discrete values. Any variable which can take only a finite number of different values is, of course, a discrete variable. A variable which can take infinitely many values may still be discrete. For example, if the variable x is restricted to integral values, x is discrete. The c.d.f. of a discrete random variable is a step function, as shown in Fig. 2. Thus, a discrete random variable has no probability density function, but admits an elementary probability law $f(t)$, where $f(t)$ denotes the probability that $x = t$.

In what follows we shall consider only random variables which either admit a probability density function or have a discrete distribution. By the probability distribution, or more briefly distribution, $f(t)$, of a random variable x, we shall always mean the probability density function of x, if a probability density function exists. If x is a discrete random variable, $f(t)$ will denote the probability that $x = t$. We shall sometimes refer to the distribution $f(t)$ of x also as the population distribution of x, or the distribution of x in the population.

1.1.5 Expected Value and Higher Moments of a Random Variable

Suppose that x is a random variable which has a discrete distribution. Let $f(t)$ denote the distribution of x, i.e., $f(t)$ is the probability that $x = t$. Then the expected value of x, in symbols $E(x)$, is defined by

$$(1:4) \qquad E(x) = \sum_t t\, f(t)$$

where the summation is to be taken over all possible values t of x. Interpreting the probability $f(t)$ as the proportion of objects in the population for which $x = t$, we see from (1:4) that the expected value $E(x)$ of x is the same as the mean value of x in the population. If x is a continuous variable which admits a probability density function $f(t)$, then the expected value of x is given by

$$(1:5) \qquad E(x) = \int_{-\infty}^{+\infty} t\, f(t)\, dt$$

The expected value of x is often called also the population mean, or mean of x.

A function $\phi(x)$ of a random variable x is itself a random variable. For any positive integer r and for any constant c, the expected value of $(x - c)^r$ is called the rth population moment of x referred to the value c. Of special interest is the case in which $c = E(x)$. The expected value of $[x - E(x)]^r$ is called the rth moment of x referred to the mean. The second moment referred to the mean, i.e., the expected value of $[x - E(x)]^2$, is also called the variance of x. The square root of the variance is called the standard deviation.

Consider the normal probability density function

$$(1:6) \qquad f(t) = \frac{1}{\sqrt{2\pi}\sigma} e^{-\frac{1}{2\sigma^2}(t-\mu)^2}$$

where μ and σ are constants ($\sigma > 0$). Let x be a random variable whose distribution is given by (1:6). That the expected value of x is then equal to μ and the variance of x is equal to σ^2 can easily be verified.

1.2 Notion of a Statistical Hypothesis

1.2.1 Unknown Parameters of a Distribution

Let x be a random variable. A statistical problem arises when the distribution of x is not known and we want to draw some inference concerning the unknown distribution of x on the basis of a limited number of observations on x. Frequently, the distribution of x is not entirely unknown, i.e., some partial knowledge of the distribution of x is available a priori. To illustrate this we shall consider the two following examples.

Example 1. Consider a lot consisting of N units of a certain manufactured product. Suppose that each unit is classified in one of the two categories, defective and non-defective. The value 0 is assigned to each non-defective unit and the value 1 to each defective unit. One unit is drawn at random from the lot and is inspected. The outcome x of this experiment is a random variable which can take only the values 0 and 1. Denote by p the proportion of defectives in the lot. Then the probability that $x = 1$ is equal to p and the probability that $x = 0$ is equal to $1 - p$. Thus, if the value of p were known, the distribution of x would be completely known. Usually p is unknown and we want to make some inference regarding the value of p by inspecting a limited number of units drawn from the lot. If p is unknown, we have only partial knowledge of the distribution of x; we know merely that x is restricted to the values 0 and 1. In this case p is considered an unknown parameter which can have any value between 0 and 1. We shall also say that the distribution of x involves an unknown parameter p. Thus in this example the distribution of x is known except for the value of an unknown parameter p.

Example 2. Suppose that the length of a bar is measured with an instrument for which the error of measurement is known to be normally distributed. The outcome x of such a measurement is then a normally distributed random variable, i.e., the distribution of x is given by the normal density function

$$\frac{1}{\sqrt{2\pi}\sigma} e^{-\frac{1}{2\sigma^2}(x-\mu)^2}$$

Usually the mean μ and the variance σ^2 of the distribution are unknown. These quantities are also called the parameters of the normal distribution. The mean μ can take any real value and σ^2 can take any positive value. Thus, in this example too, the distribution function is known except for the values of the parameters μ and σ^2 involved in the distribution function.

A general situation similar to that given in Examples 1 and 2 may be described as follows: *The functional form of the distribution function is known and merely the values of a finite number of parameters involved in the distribution function are unknown; i.e., the distribution function is known except for the values of a finite number of parameters.* In Example 1 the only unknown parameter is the proportion p of defectives in the lot. In Example 2 there are two unknown parameters, the mean μ and the variance σ^2.

In what follows we shall assume that the distribution of the random variable x is known except for the values of a finite number of parameters.

1.2.2 Simple and Composite Hypotheses

Let $\theta_1, \cdots, \theta_k$ be the unknown parameters of the distribution of the random variable x under consideration. A statement about the values of $\theta_1, \cdots, \theta_k$ is called a *simple hypothesis* if it determines uniquely the values of all k parameters. It is called a *composite hypothesis* if it is consistent with more than one value for some parameter. For example, if there are two unknown parameters, θ_1 and θ_2, involved in the distribution of x, the hypothesis that $\theta_1 = 2$ and $\theta_2 = 4$ is a simple hypothesis, since it specifies completely the values of the unknown parameters. On the other hand, the hypothesis that $\theta_1 = \theta_2$ is composite. In Example 1 the statement that the unknown proportion p of defectives is equal to .2 is a simple hypothesis. On the other hand, the statement that p lies between .1 and .3 is a composite hypothesis. In Example 2 the statement that $\mu = 3$ would be a composite hypothesis, since it does not specify the value of the unknown variance σ^2.

In general, the parameters $\theta_1, \cdots, \theta_k$ will not be subject to any a priori restrictions; i.e., they may take any values. However, the parameters may in some cases be restricted to certain intervals. For instance, if one of the unknown parameters is the standard deviation, this parameter is restricted to positive values. In other cases, the parameter may be able to take only a finite number of discrete values.

1.3 Outline of the Current Procedure for Testing Statistical Hypotheses

1.3.1 The Sample

Let x be a random variable and suppose that we wish to test a hypothesis concerning the unknown parameters of the distribution of x. The decision to accept or reject the hypothesis in question is always made on the basis of a finite number of observations on x. A set of a finite number of observations on x is called a sample. The number of observations contained in the sample is called the size of the sample.

We shall be concerned mostly with the case in which the successive observations on x are independent in the probability sense. The successive observations $x_1, \cdots, x_n$ on x are said to be independent in the probability sense if the (conditional) probability distribution of the ith observation x_i $(i = 2, \cdots, n)$, when the values of the preceding observations $x_1, \cdots, x_{i-1}$ are known, is not affected by these values. This condition cannot be strictly fulfilled if the successive observations are drawn from a finite population. Consider, for instance, the case discussed in Example 1 on page 12. Suppose that two successive units are drawn at random from the lot. Denote by x_1 the value of x for the first unit and by x_2 the value of x for the second unit. The distribution of x_1 is clearly given as follows: the probability that $x_1 = 0$ is $1 - p$ and the probability that $x_1 = 1$ is equal to p. The distribution of x_2, when the value of x_1 is known, is given as follows: if $x_1 = 0$, then the probability that $x_2 = 1$ is equal to $pN/(N - 1)$ and the probability that $x_2 = 0$ is equal to $1 - [pN/(N - 1)]$. On the other hand, if $x_1 = 1$, the probability that $x_2 = 1$ is equal to $(pN - 1)/(N - 1)$ and the probability that $x_2 = 0$ is equal to $1 - [(pN - 1)/(N - 1)]$. Thus, the probability distribution of x_2 is affected by the outcome of x_1. For similar reasons no strict independence can prevail in any other case in which the successive observations are drawn from a finite population. However, if the number of objects in the finite population is sufficiently large, the dependence is only slight and can be neglected.

Let x be a discrete random variable, and denote the distribution of x by $f(t)$, i.e., $f(t)$ is the probability that $x = t$. Let $x_1, \cdots, x_n$ be a set of n independent observations on x. Because of the independence of the observations, the probability of obtaining a sample equal to the observed one is given by the product

$$f(x_1)f(x_2) \cdots f(x_n)$$

This product is also called the joint probability distribution of the observations $x_1, \cdots, x_n$.

If x is a continuous random variable admitting a probability density function $f(x)$, then the joint density function of n independent observations $x_1, \cdots, x_n$ on x is given by the product

$$f(x_1)f(x_2) \cdots f(x_n)$$

1.3.2   The General Nature of a Test Procedure

Denote by n the number of observations on the basis of which the acceptance or rejection of the hypothesis in question is to be decided. Any possible outcome of n successive observations is a sample of size n. A test procedure leading to the acceptance or rejection of the hypothesis in question is simply a rule specifying, for each possible sample of size n, whether the hypothesis should be rejected or accepted on the basis of that sample. This may also be expressed as follows: A test procedure is simply a subdivision of the totality of all possible samples of size n into two mutually exclusive parts, say part 1 and part 2, together with the application of the rule that the hypothesis be rejected if the observed sample is contained in part 1 and that the hypothesis be accepted if the observed sample is contained in part 2. Part 1 is also called the critical region. Since part 2 is the totality of all samples of size n which are not included in part 1, part 2 is uniquely determined by part 1. Thus, choosing a test procedure is equivalent to determining a critical region.

As an illustration, we shall discuss a few examples. Suppose that a lot consisting of N units of a manufactured product is submitted for acceptance inspection. Assume that each unit is classified in one of the two categories: defective and non-defective. The proportion p of defectives in the lot is assumed to be unknown. Let p_0 be a value between 0 and 1 such that we prefer to accept the lot if the proportion p of defectives is $\leq p_0$ and we prefer to reject the lot if $p > p_0$. Suppose that a sample of n units, drawn at random from the lot, is inspected and on the basis of this sample a decision is to be made to accept the lot or reject it. In other words, on the basis of the inspection of the sample of n units a decision is to be made to accept the hypothesis $p \leq p_0$ or reject it. The critical region generally used in this case is defined as follows: The hypothesis that $p \leq p_0$ is rejected, i.e., the lot is rejected, if, and only if, the proportion of defectives in the observed sample of n units exceeds a suitably chosen numerical constant c.

Another example: Suppose that the length of a bar is measured with an instrument for which the error of measurement is known to be normally distributed with variance equal to unity. Thus, the outcome x of a measurement is a normally distributed random variable with mean μ equal to the true length of the bar and variance unity. Let the hypothesis to be tested be the statement that the true length of the bar is equal to a specified value μ_0. This hypothesis is to be tested on the basis of a sample consisting of n independent measurements $x_1, \cdots, x_n$ on the length of the bar. The critical region generally used for this purpose is defined as follows: The hypothesis that $\mu = \mu_0$ is rejected if, and only if, the sample observed is such that $|\bar{x} - \mu_0| \geq c$ where $\bar{x}$ denotes the arithmetic mean of the n observations and c is a suitably chosen numerical constant.

There are, in general, infinitely many possibilities for choosing a critical region. For instance, in the example just discussed we could have used the median, or the geometric mean, or the harmonic mean, or some other mean of the observations instead of the arithmetic mean. The various critical regions cannot be regarded as equally good and the fundamental problem in testing hypotheses is to set up principles for the proper choice of the critical region. Such principles have been advanced by Jerzy Neyman and Egon S. Pearson. In the next section we shall discuss briefly the basic idea of the Neyman-Pearson theory.[3]

[3] See, for example, J. Neyman and E. S. Pearson, *Statistical Research Memoirs*, University College, London, Vol. I (1936), pp. 1–37.

1.3.3   Principles for Choosing a Critical Region

The principles formulated by Neyman and Pearson for the proper choice of a critical region constituted an advance of fundamental importance in the theory of testing hypotheses. The purpose of this section is to indicate briefly the basic idea of the Neyman-Pearson theory.

A simple case of particular theoretical interest arises when only one unknown parameter θ is involved in the distribution of the random variable x under consideration, and θ can take only two values, θ_0 and θ_1. The basic idea of the Neyman-Pearson theory can be indicated even in this simple case. Therefore, in the rest of this section, as well as in the following section, 1.3.4, we shall restrict ourselves to the case of a single parameter θ which can take only two values, θ_0 and θ_1.

For any value θ of the parameter, let $f(x, \theta)$ denote the distribution of x. We shall denote $f(x, \theta_0)$ by $f_0(x)$ and $f(x, \theta_1)$ by $f_1(x)$. Suppose that it is desired to test the hypothesis that $\theta = \theta_0$. We shall refer to this hypothesis as the null hypothesis and denote it by H_0. The hypothesis that $\theta = \theta_1$ will be called the alternative hypothesis and will be denoted by H_1. Thus, we shall deal with the problem of testing the hypothesis H_0 against the alternative hypothesis H_1 on the basis of a sample of n independent observations $x_1, \cdots, x_n$ on x.

As a basis for choosing among critical regions the following considerations have been advanced by Neyman and Pearson: In accepting or rejecting H_0, we may commit errors of two kinds. We commit an error of the first kind if we reject H_0 when it is true; we commit an error of the second kind if we accept H_0 when H_1 is true. After a particular critical region W has been chosen, the probability of committing an error of the first kind, as well as the probability of committing an error of the second kind, is uniquely determined. The probability of committing an error of the first kind is equal to the probability, determined on the assumption that H_0 is true, that the observed sample will be included in the critical region W. The probability of committing an error of the second kind is equal to the probability, determined on the assumption that H_1 is true, that the observed sample will fall outside the critical region W. For any given critical region W we shall denote the probability of an error of the first kind by α and the probability of an error of the second kind by β.

The probabilities α and β have the following important practical interpretation: Suppose we draw a large number of samples of size n. Let M be the number of such samples drawn. Suppose that for each of these M samples we reject H_0 if the sample is included in W and accept H_0 if the sample lies outside W. In this way we make M statements of rejection or acceptance. Some of these statements will in general be wrong. If H_0 is true and if M is large, the probability is nearly 1 (i.e., it is practically certain) that the proportion of wrong statements (i.e., the number of wrong statements divided by M) will be approximately α. If H_1 is true, the probability is nearly 1 that the proportion of wrong statements will be approximately β. Thus, we can say that in the long run the proportion of wrong statements will be α if H_0 is true and β if H_1 is true.

It is clear that one critical region W is more desirable than another if it has smaller values of α and β. Although either α or β can be made arbitrarily small by a proper choice of the critical region W, it is impossible to make both α and β arbitrarily small for a fixed value of n, i.e., a fixed sample size. To illustrate this point, consider the following two extreme cases: (1) W is empty, i.e., we always accept H_0, irrespective of the outcome of the sample. In this case $\alpha = 0$ and $\beta = 1$. (2) W is the totality of all possible samples, i.e., we always reject H_0. In this case $\alpha = 1$ and $\beta = 0$. If, for some reason, we decide to consider only critical regions W for which α has a given fixed value, the choice of W is based on the following principle, introduced by Neyman and Pearson: Restricting ourselves to regions W for which α has a fixed value, we choose that one for which β is a minimum.

The quantity α is called the size of the critical region, and the quantity $1 - \beta$, the power of the critical region. A critical region which has the highest power in the class of all regions of equal size is a most powerful region. Since minimizing β is the same as maximizing $1 - \beta$, the Neyman-Pearson principle concerning the choice of the critical region W can be formulated as follows: Restricting ourselves to regions of a fixed size α, we choose that one which is most powerful.

For a fixed sample size, the probability β is a (single-valued) function of α, say $\beta(\alpha)$, if a most powerful critical region is used. Thus, given the number of observations on which the test is based, one of the quantities α and β can still be chosen arbitrarily. The Neyman-Pearson theory leaves the question of this choice open. It is clear that if α is small, $\beta(\alpha)$ is in general large, and if α is large, $\beta(\alpha)$ is in general small. The choice of α (or β) will be greatly influenced by the relative importance of the errors of the first and second kinds in each particular application. Suppose, for example, that the loss caused by an error of the first kind is one dollar and the loss caused by an error of the second kind is merely one cent. Then a small α and a large β will be preferable to a large α and a small β.

Neyman and Pearson show that a region consisting of all samples $(x_1, \cdots, x_n)$ which satisfy the inequality

$$(1:7) \qquad \frac{f_1(x_1)f_1(x_2) \cdots f_1(x_n)}{f_0(x_1)f_0(x_2) \cdots f_0(x_n)} \geq k$$

is a most powerful critical region for testing the hypothesis H_0 against the alternative hypothesis H_1. The term k on the right-hand side of (1:7) is a constant chosen so that the region will have the required size α. The reason why the critical region defined by (1:7) is most powerful can be indicated as follows: For simplicity suppose that the probability distributions under H_0 and H_1 are discrete. Thus, $f_i(x_1)f_i(x_2) \cdots f_i(x_n)$ $(i = 0, 1)$ denotes the probability of obtaining a sample equal to the observed one. The critical region defined by (1:7) can be built up by starting with a sample $E^1 = (x_1{}^1, x_2{}^1, \cdots, x_n{}^1)$ for which $\dfrac{f_1(x_1) \cdots f_1(x_n)}{f_0(x_1) \cdots f_0(x_n)}$ takes its maximum value. Then a sample $E^2 = (x_1{}^2, \cdots, x_n{}^2)$ is included for which $\dfrac{f_1(x_1) \cdots f_1(x_n)}{f_0(x_1) \cdots f_0(x_n)}$ takes its maximum value in the set of samples which is left after E^1 has been removed from the totality of all possible samples. In general, after r samples $E^1, \cdots, E^r$ have been included in the critical region, a sample E^{r+1} is added for which $\dfrac{f_1(x_1) \cdots f_1(x_n)}{f_0(x_1) \cdots f_0(x_n)}$ takes its maximum value in the set of samples $(x_1, \cdots, x_n)$ which are left after $E^1, \cdots, E^r$ have been removed from the totality of all samples. This construction is continued until the size of the region reaches the desired value α.[4] Since at any stage of the construction the last sample included in the critical region has the largest probability under H_1 per unit probability under H_0 as compared with any other sample not yet included in the region, it can be seen that the probability measure of the critical region under H_1, i.e., the power of the critical region, is greater than or equal to the power of any other region of equal size.

Let us illustrate the principle for choosing a critical region by application to a simple and familiar case. Let H_0 be the hypothesis that x is normally distributed with mean θ_0 and variance unity. Let H_1 be the hypothesis that x is normally distributed with mean θ_1 and vari-

---

[4] If x is a discrete variable, it may happen that, at the last stage of the construction, at the inclusion of the last sample in the critical region, the size of the region increases from a value below α to a value somewhat greater than α.

ance unity. Assume $\theta_1 > \theta_0$. For testing H_0 against H_1 we shall have to determine the ratio $\dfrac{f_1(x_1) \cdots f_1(x_n)}{f_0(x_1) \cdots f_0(x_n)}$. Since

$$f_1(x_1) \cdots f_1(x_n) = \frac{1}{(2\pi)^{\frac{n}{2}}} e^{-\frac{1}{2} \sum_{\alpha=1}^{n} (x_\alpha - \theta_1)^2}$$

and

$$f_0(x_1) \cdots f_0(x_n) = \frac{1}{(2\pi)^{\frac{n}{2}}} e^{-\frac{1}{2} \sum_{\alpha=1}^{n} (x_\alpha - \theta_0)^2}$$

the inequality (1:7) can be written as

$$(1:8) \qquad \frac{e^{-\frac{1}{2} \sum_{\alpha=1}^{n} (x_\alpha - \theta_1)^2}}{e^{-\frac{1}{2} \sum_{\alpha=1}^{n} (x_\alpha - \theta_0)^2}} \geqq k$$

Taking the logarithm on both sides of this inequality, we obtain

$$\tfrac{1}{2}\Sigma(x_\alpha - \theta_0)^2 - \tfrac{1}{2}\Sigma(x_\alpha - \theta_1)^2 = (\theta_1 - \theta_0)\Sigma x_\alpha + \tfrac{1}{2}n(\theta_0{}^2 - \theta_1{}^2) \geqq \log k$$

Hence

$$(1:9) \qquad \sum_{\alpha=1}^{n} x_\alpha \geqq \frac{\log k - \tfrac{1}{2}n(\theta_0{}^2 - \theta_1{}^2)}{\theta_1 - \theta_0} = k' \quad \text{(say)}$$

Inequality (1:9) can be written as

$$(1:10) \qquad \frac{\Sigma(x_\alpha - \theta_0)}{n} \geqq \frac{k' - n\theta_0}{n} = k'' \quad \text{(say)}$$

Now we shall determine the value of k'' such that the critical region defined by the inequality (1:10) has the size $\alpha = .05$. Since under the hypothesis H_0 the random variable $[\Sigma(x_\alpha - \theta_0)]/n$ is normally distributed with zero mean and variance $1/n$, we see from a table of the normal distribution that $k'' = 1.64/\sqrt{n}$. Thus, the most powerful region of size .05 consists of all samples for which the inequality

$$(1:11) \qquad \frac{\Sigma(x_\alpha - \theta_0)}{n} \geqq \frac{1.64}{\sqrt{n}}$$

holds.

This is a familiar result. Long before Neyman and Pearson developed their theory of testing hypotheses, it had been the practice to use the critical region (1:11) for testing the hypothesis that $\theta = \theta_0$

against alternative values $\theta > \theta_0$. A remarkable feature of the region given by (1:11) is that it does not depend on the alternative value θ_1. In the derivation of (1:11) merely the inequality $\theta_1 > \theta_0$ was used. Hence, the test defined by the region (1:11) is most powerful with respect to *all* alternatives $\theta > \theta_0$, i.e., it is a *uniformly most powerful* test when the alternatives are restricted to values greater than θ_0.

1.3.4 Number of Observations Necessary if α and β Have Pre-assigned Values

In the preceding section we assumed that α and the sample size n were given and we were looking for a critical region for which β was a minimum. In this section we shall assume that α and β are given and our problem is to determine the *minimum* value of n for which the power of the most powerful region of size α is greater than or equal to $1 - \beta$.

Let β_n denote the probability of an error of the second kind associated with a most powerful critical region of size α when the test is based on n observations. It can be shown that β_n decreases, or at least does not increase, with increasing n. In general, β_n will approach 0 as n increases indefinitely. Denote by $n(\alpha, \beta)$ the smallest value of n for which $\beta_n \leq \beta$. If we want a test procedure such that the probability of an error of the first kind is equal to α and the probability of an error of the second kind does not exceed β, then according to the current theory we must draw a sample of size $n \geq n(\alpha, \beta)$. If we use a most powerful critical region, we need a sample of size $n = n(\alpha, \beta)$.

1.3.5 Testing a Hypothesis Viewed as a Decision between Two Courses of Action

It happens frequently in practice that we have to decide between two courses of action, say action 1 and action 2, and the preference for one or the other action depends on the value of an unknown parameter θ of the distribution of a random variable x. Denote by ω the set of all values of θ for which action 2 is not preferable to action 1. Thus, for any value θ not contained in ω we prefer action 2 to action 1. The problem of deciding between these two actions on the basis of a sample of n independent observations on x may be formulated as a problem of testing the hypothesis H that the true value of θ is contained in the set ω. If the test procedure leads to the acceptance of H we take action 1, and if it leads to the rejection of H we take action 2.

Consider, for example, the following problem. A lot consisting of a large number of units of a manufactured product is submitted for acceptance inspection. Suppose that the proportion p of defectives in the lot is unknown. There are two courses of action: acceptance of the lot and rejection of the lot. In general, there will exist a particular value p' of p such that if the true proportion of defectives is $< p'$ we prefer acceptance and if $p > p'$ we prefer rejection. If $p = p'$ we are indifferent which action is taken. Suppose that a decision is to be made on the basis of a sample of n units drawn at random from the lot. This problem may be viewed as a problem of testing the hypothesis H that $p \leq p'$ on the basis of a sample drawn from the lot. The lot is accepted or rejected according as H is accepted or rejected.

As mentioned in Section 1.3.3, the choice of α, i.e., the size of the critical region, is greatly influenced by the relative importance we attach to errors of the first and second kinds. If the problem of testing a hypothesis arises out of the problem of deciding between certain two courses of action, the relative importance of the errors of the first and second kinds may be judged by considering the practical consequences of taking one action when the value of the parameter is such that the other action would have been preferable.

Chapter 2. SEQUENTIAL TEST OF A STATISTICAL HYPOTHESIS: GENERAL DISCUSSION

2.1 Notion of a Sequential Test

In the current theory of testing hypotheses the number of observations, i.e., the size of the sample on which the test is based, is treated as a constant for any particular problem. An essential feature of the sequential test, as distinguished from the current test procedure, is that the number of observations required by the sequential test depends on the outcome of the observations and is, therefore, not predetermined, but a random variable.

The sequential method of testing a hypothesis H may be described as follows. A rule is given for making one of the following three decisions at any stage of the experiment (at the mth trial for each integral value of m): (1) to accept the hypothesis H, (2) to reject the hypothesis H, (3) to continue the experiment by making an additional observation. Thus, such a test procedure is carried out sequentially. On the basis of the first observation one of the aforementioned three decisions is made. If the first or second decision is made, the process is terminated. If the third decision is made, a second trial is performed. Again, on the basis of the first two observations one of the three decisions is made. If the third decision is made, a third trial is performed, and so on. The process is continued until either the first or the second decision is made. The number n of observations required by such a test procedure is a random variable, since the value of n depends on the outcome of the observations.

For each positive integral value m, we shall denote by M_m the totality of all possible samples $(x_1, \cdots, x_m)$ of size m. We shall also refer to M_m as the m-dimensional sample space. A rule for making one of the three decisions at any stage of the experiment can be described as follows. For each integral value m, the m-dimensional sample space is split into three mutually exclusive parts, $R_m{}^0$, $R_m{}^1$, and R_m. After the first observation x_1 has been drawn, the hypothesis H that is being tested is accepted if x_1 lies in $R_1{}^0$; H is rejected if x_1 lies in $R_1{}^1$; or a second observation is made if x_1 lies in R_1. If the third decision is made and a second observation x_2 drawn, H is accepted, H is rejected, or a third observation is drawn, according as the observed sample (x_1, x_2) lies in $R_2{}^0$, $R_2{}^1$, or R_2. If (x_1, x_2) lies in R_2, a third observation x_3 is drawn and one of the three decisions is made according as (x_1, x_2, x_3) lies in $R_3{}^0$, $R_3{}^1$, or R_3, and so on. This process is stopped when, and only when, either the first or the second decision is made.[1] Thus, a sequential test is completely defined by defining the sets $R_m{}^0$, $R_m{}^1$, and R_m for all positive integral values m. Since $R_m{}^0$, $R_m{}^1$, and R_m are mutually exclusive and add up to the whole sample space M_m, it is sufficient to define any two of the sets $R_m{}^0$, $R_m{}^1$, and R_m. Any one of the three sets $R_m{}^0$, $R_m{}^1$, and R_m consists precisely of all those samples which are not contained in the other two.

We shall call a sample $(x_1, \cdots, x_m)$ ineffective if it contains an initial segment $(x_1, \cdots, x_{m'})$, where $m' < m$, such that $(x_1, \cdots, x_{m'})$ lies in $R_{m'}{}^0$ or in $R_{m'}{}^1$. A sample which is not ineffective will be said to be an effective sample. Clearly, for a sequential test procedure we shall have an effective sample at any stage of the experiment. Thus, in defining the sets $R_m{}^0$, $R_m{}^1$, and R_m we may disregard ineffective samples. In other words, it is sufficient to state in which of the sets $R_m{}^0$, $R_m{}^1$, and R_m each effective sample $(x_1, \cdots, x_m)$ should be included, since ineffective samples cannot occur during the sequential process.

The following is a simple example of a sequential test. Suppose that a lot consisting of a large number of units of a manufactured product is submitted for acceptance inspection. Each unit is classified in one of the two categories: defective and non-defective. The proportion p of defectives in the lot is unknown. The lot is considered acceptable if $p \leq$ a given value p'. If $p > p'$ we prefer to reject the lot. Thus, we are interested in testing the hypothesis H that $p \leq p'$. The following procedure of testing H is a simple example of a sequential test. Let n_0 denote a given integer. If the first n_0 units inspected are non-defective, we stop inspection and the lot is accepted (H is accepted). If for some value $m \leq n_0$ the mth unit inspected is found defective, no further units are inspected and the lot is rejected (H is rejected). We shall assign the value 0 to any non-defective unit and the value 1 to any defective unit. In this example, a sample $(x_1, \cdots, x_m)$ is effective if and only if $m \leq n_0$ and $x_1 = \cdots = x_{m-1} = 0$. $R_m{}^0$ contains no effective sample for $m < n_0$, i.e., acceptance is not possible for $m < n_0$. $R_{n_0}{}^0$ contains only one effective sample: $(0, 0, \cdots, 0)$. For any $m \leq n_0$ the set $R_m{}^1$ contains exactly one effective sample: $(0, 0, \cdots, 0, 1)$.

The sets $R_m{}^0$, $R_m{}^1$, and R_m ($m = 1, 2, \cdots$) defining a sequential test can be chosen in many ways, and a fundamental problem in the theory of sequential tests is that of a proper choice of these sets. To formulate

---

[1] We shall consider only sequential tests for which the probability is one that the process will eventually terminate.

principles for a proper choice of the sets $R_m{}^0$, $R_m{}^1$, and R_m, it is necessary to study the consequences of any particular choice. This will be done in the next section.

2.2 Consequences of the Choice of Any Particular Sequential Test

2.2.1 The *Operating Characteristic* Function

After a particular sequential test has been adopted, i.e., a particular choice of the sets $R_m{}^0$, $R_m{}^1$, and R_m $(m = 1, 2, \cdots)$ has been made, the probability that the process will terminate with the acceptance of the hypothesis H_0 under test depends only on the distribution of the random variable x under consideration. As before, it is assumed that the distribution of x is known except for the values of a finite number of parameters, $\theta_1, \cdots, \theta_k$, say. Thus, the distribution of x is given by a function $f(x, \theta_1, \cdots \theta_k)$ where the functional form f is known, but the true values of the parameters $\theta_1, \cdots, \theta_k$ are unknown. To simplify notation, we shall use the letter θ without subscript to denote the set of all k parameters $\theta_1, \cdots, \theta_k$. We shall refer to θ as a parameter point, since θ can be represented geometrically by a point with the coordinates $\theta_1, \cdots, \theta_k$. Since the distribution of x is determined by the parameter point θ, the probability of accepting H_0 will be a function of θ. This function will be denoted by $L(\theta)$ and will be called the *operating characteristic* (OC) function. If there is only one unknown parameter θ the function $L(\theta)$ can be plotted as a curve, θ being measured along the horizontal axis and $L(\theta)$ along the vertical axis. Since we shall consider only tests for which the probability that the procedure will eventually terminate is equal to 1, the probability of rejecting H_0 is equal to $1 - L(\theta)$.

The OC function is very closely related to the notion of the power function in the current theory of tests. For any parameter point θ which is not consistent with the null hypothesis H_0, the power of the test is defined as the probability of rejecting H_0 when θ is the true point. Thus, for any θ not consistent with H_0 the power of the test is equal to $1 - L(\theta)$.

To illustrate the meaning of an OC function, we shall compute the OC function of the particular sequential test given as an example in the preceding section. In that example the only unknown parameter is $\theta = p$, where p denotes the proportion of defectives in the lot. The lot is accepted if, and only if, the first n_0 units inspected are non-defective. The probability that the first unit inspected is non-defective is equal to $1 - p$. On the assumption that the size of the lot is sufficiently large as compared with n_0, the successive observations may be treated as being independent. Then the probability that all n_0 units will be non-defective is equal to $(1 - p)^{n_0}$. Thus, the operating characteristic function is given by

$$L(p) = (1 - p)^{n_0}$$

This function can be plotted, as shown in Fig. 4, by measuring p along the horizontal axis and $L(p)$ along the vertical axis.

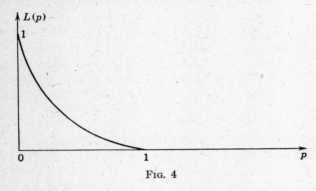

FIG. 4

The OC function describes what the sequential test procedure accomplishes. For any parameter point θ the probability of making a correct decision can be obtained immediately from the OC function. If the parameter point θ is consistent with the hypothesis H_0 to be tested, then the probability of making a correct decision is equal to $L(\theta)$. If the true parameter point θ is not consistent with the hypothesis H_0, the probability of making a correct decision is equal to $1 - L(\theta)$. Clearly, an OC function is considered more favorable the higher the value of $L(\theta)$ for θ consistent with H_0 and the lower the value of $L(\theta)$ for θ not consistent with H_0.

2.2.2 The Average (Expected) Sample Number (ASN) Function of a Sequential Test

We have pointed out before that the number of observations required by a sequential test is not predetermined, but is a random variable, because at any stage of the experiment the decision to terminate the process depends on the results of the observations made so far. For example, for the particular sequential test discussed in the preceding section, the number of observations required by the test may be anything from 1 to n_0. If no defects are found during the sampling process, we shall make n_0 observations. On the other hand, if the first $m - 1$ units inspected are non-defective and the mth unit is defective for some value $m < n_0$, then the total number of observations made will be equal to m.

We shall denote by n the number of observations required by the sequential test. Then n is a random variable. Carrying out the same sequential test procedure repeatedly, we shall obtain, in general, different values for n. Of particular interest is the expected value of n (the average value of n in the long run, when the same test procedure is applied repeatedly). For any given test procedure the expected value of n depends only on the distribution of x. Since the distribution of x is determined by the parameter point θ, the expected value of n depends only on θ. For any given parameter point θ, we shall denote the expected value of n by $E_\theta(n)$. If there is only one unknown parameter θ the function $E_\theta(n)$ can be plotted as a curve, θ being measured along the horizontal axis and $E_\theta(n)$ along the vertical axis. We shall refer to the average sample number function $E_\theta(n)$ briefly as the ASN function.

As an example, we shall compute the ASN function for the particular sequential test discussed in the preceding section. For any positive integral value $m < n_0$, the probability that the test will be terminated at the mth observation is given by $(1 - p)^{m-1}p$. We shall inspect n_0 units if and only if the first $n_0 - 1$ units are found non-defective. Thus, the probability that the test will require exactly n_0 observations is equal to $(1 - p)^{n_0-1}$. Hence, the expected value of n is given by

$$E_p(n) = \sum_{m=1}^{n_0-1} mp(1 - p)^{m-1} + n_0(1 - p)^{n_0-1}$$

The graph of the ASN function will be of the type shown in Fig. 5.

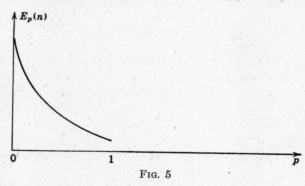

FIG. 5

An OC function and an ASN function are associated with each test procedure. These two functions are perhaps the most important consequences of a test procedure. The OC function describes how well the test procedure achieves its objective of making correct decisions, and the ASN function represents the price we have to pay, in terms of the number of observations required for the test. Thus, in judging the relative merits of two different test procedures, we shall compare the OC and ASN functions of these two tests.

2.3 Principles for the Selection of a Sequential Test

2.3.1 Degree of Preference for Acceptance or Rejection of the Null Hypothesis H_0 as a Function of the Parameter θ

In order to set up principles for the selection of a sequential test it is necessary to investigate the dependence of the preference for rejection or acceptance of the null hypothesis H_0 on the parameter point θ. Denote by ω the set of all those parameter points θ which are consistent with H_0, i.e., H_0 is precisely the statement that the true parameter point is included in the set ω. For example, if there is only one unknown parameter θ and if H_0 is the hypothesis that θ is less than or equal to a certain particular value θ_0, ω is the set of all values θ for which $\theta \leq \theta_0$. Since a correct decision is preferred to a wrong decision, we can say that acceptance of H_0 is preferred whenever θ is in ω, and rejection of H_0 is preferred whenever θ is outside ω.

The mere statement of preference for acceptance or rejection of H_0 is not yet a sufficient guide for the selection of a proper sequential test. For this purpose it is necessary to know something about the *degree* of preference for acceptance or rejection as a function of the parameter point θ.

We shall denote by $\bar{\omega}$ the set of all parameter points which lie outside ω. A point θ will be said to be on the boundary of ω, or a boundary point of ω, if any arbitrarily small neighborhood of θ contains points of ω as well as of $\bar{\omega}$. The totality of all boundary points of ω will be called the boundary of ω. If, for example, there is only one unknown parameter and ω is defined by $\theta \leq \theta_0$, then θ_0 is the only boundary point of ω. If ω is the set of all values θ for which $\theta_0 \leq \theta \leq \theta_1$, then both θ_0 and θ_1 are boundary points. If the true parameter point θ lies in ω but is near the boundary of ω, the preference for acceptance of H_0 will, in general, be only slight. Similarly, if the true point θ lies in $\bar{\omega}$ but near the boundary of ω, the preference for rejection of H_0 will be only slight. In other words, the rejection of H_0 is not considered to be a serious error if θ is in ω but near the boundary. Similarly, the acceptance of H_0 is not considered a serious error if θ is in $\bar{\omega}$ but near the boundary of ω. If the true point θ lies exactly on the boundary of ω, there will be, in general, no definite preference for one or the other action, i.e., it will be indifferent to us whether the hypothesis H_0 is accepted or rejected.

In general, it will be possible to subdivide the totality of all parameter points (parameter space) into three mutually exclusive zones: a zone consisting of all points θ for which acceptance of H_0 is strongly preferred; a zone consisting of points θ for which rejection of H_0 is strongly preferred; and a zone consisting of all points θ which are not included in either of the first two zones, i.e., the third zone consists of all points θ for which neither acceptance nor rejection of H_0 is strongly preferred. We shall refer to the first zone as the zone of preference for acceptance, to the second zone as the zone of preference for rejection, and to the third zone as the zone of indifference. The zone of preference for acceptance will always be a subset of ω and the zone of preference for rejection will be a subset of $\bar{\omega}$. The zone of indifference will usually consist of points of ω and $\bar{\omega}$ which are near the boundary or on the boundary of ω.

Although the subdivision of the parameter space into three zones as described above is used as a basis for the selection of a sequential test, it cannot be considered a statistical problem. Such a subdivision is made in each case on the basis of practical considerations concerning the consequences of a wrong decision.

The subdivision of the parameter space into the above-mentioned three zones gives a somewhat sketchy picture of the degree of preference for acceptance or rejection as a function of the parameter θ. A more refined description of the degree of preference for one or the other action can be given in terms of two functions $w_0(\theta)$ and $w_1(\theta)$, where $w_0(\theta)$ expresses the relative importance of, i.e., the loss caused by, the error of accepting H_0 when θ is true, and $w_1(\theta)$ expresses the relative importance of the error of rejecting H_0 when θ is true. The function $w_0(\theta) = 0$ for any θ in ω, since for such points θ the acceptance of H_0 is a correct decision. For any θ in $\bar{\omega}$, $w_0(\theta)$ will have a positive value which will, in general, increase with increasing distance of θ from the boundary of ω. Similarly, $w_1(\theta) = 0$ for all θ in $\bar{\omega}$ and $w_1(\theta) > 0$ for all θ in ω. Again, $w_1(\theta)$ will, in general, increase with increasing distance of θ from the boundary of ω. Our subdivision of the parameter space into three zones may be interpreted as being equivalent to choosing the functions $w_0(\theta)$ and $w_1(\theta)$ as follows: $w_0(\theta) = 0$ when θ is in the zone of preference for acceptance or in the zone of indifference. For any θ in the zone of preference for rejection, $w_0(\theta)$ has a high positive value, say c_0, indicating that the loss caused by acceptance is of practical importance. Similarly, $w_1(\theta) = 0$ for any

θ in the zone of preference for rejection or in the zone of indifference. For any θ in the zone of preference for acceptance, $w_1(\theta)$ has some high value, say c_1, indicating that the loss caused by rejection of H_0 is of practical importance. Although a refined description of the dependence of the degree of preference for one or the other action on θ may occasionally require the use of continuous functions $w_0(\theta)$ and $w_1(\theta)$, the step functions implied by the subdivision of the parameter space into three zones will give a sufficiently good approximation for most practical purposes. They also have the advantage of great simplicity. Thus, in what follows we shall assume that the dependence of the preference for one or the other action on θ is described by a subdivision of the parameter space into three zones of the type mentioned above.

As an illustration, we shall discuss briefly a few examples. Consider first the case in which a lot consisting of a large number of units of a manufactured product is submitted for acceptance inspection. Assuming that the units are classified in one of the two categories, defective and non-defective, the preference for acceptance or rejection of the lot depends only on the proportion p of defectives in the lot, which is unknown. In this case there is only one unknown parameter θ which is equal to the proportion p of defectives in the lot. It will, in general, be possible to select two values p_0 and p_1 $(p_0 < p_1)$ such that for any $p \leqq p_0$ the rejection of the lot is an error of practical importance, for any $p \geqq p_1$ the acceptance of the lot is considered a wrong decision of practical importance, whereas for any value p between p_0 and p_1 there is no strong preference for either action. Thus, the zone of indifference may be defined as the interval from p_0 to p_1, the zone of preference for acceptance as the set consisting of all values $p \leqq p_0$, and the zone of preference for rejection as the set of all values $p \geqq p_1$.

As a second example, consider the case in which the hardness x of a certain product varies from unit to unit such that x may be considered a normally distributed variable in the population of all units produced. Suppose that the mean value θ of x is unknown but that the standard deviation of x is known. Assume that the most desirable value of θ is θ_0 and that the product becomes less desirable as the absolute deviation $|\theta - \theta_0|$ between the true mean and the most desirable value θ_0 becomes greater. Suppose that the problem is to decide whether the product should be put on the market or not. In such a case, it will, in general, be possible to find a positive value c such that if $|\theta - \theta_0| < c$ we prefer to put the product on the market, and if $|\theta - \theta_0| > c$ we prefer to withhold the product. For $|\theta - \theta_0| = c$, we are indifferent which action is taken. Thus, the hypothesis H_0 may be defined as the hypothesis that $|\theta - \theta_0| < c$. We shall not define the zone of indifference by the equation $|\theta - \theta_0| = c$, since if $|\theta - \theta_0|$ differs only slightly from c, the preference for one action over the other is only slight and of no practical importance. However, it will be possible to find a positive value Δ such that, if $|\theta - \theta_0| < c - \Delta$, we strongly prefer to accept H_0 (to put the product on the market) and, if $|\theta - \theta_0| > c + \Delta$, we strongly prefer to reject H_0 (not to put the product on the market) whereas, if $c - \Delta \leq |\theta - \theta_0| \leq c + \Delta$, no strong preference is given to either action. Thus, the zone of indifference may be defined by the inequality $c - \Delta \leq |\theta - \theta_0| \leq c + \Delta$, the zone of preference for acceptance by $|\theta - \theta_0| < c - \Delta$, and the zone of preference for rejection by $|\theta - \theta_0| > c + \Delta$.

In each of the previous two examples there was only one unknown parameter. We shall now consider an example where there are two unknown parameters. Suppose that a lot consisting of a large number of units of a manufactured product is submitted for acceptance inspection. Assume that the characteristic of the product in which we are interested is the resistance to pressure, which is a measurable quantity x. It is assumed that x varies from unit to unit in the lot and has a normal distribution with unknown mean μ and unknown standard deviation σ. Let L be a value such that acceptance of the lot is strongly preferred if the proportion of units in the lot with resistance $x \leq L$ does not exceed .01, rejection of the lot is strongly

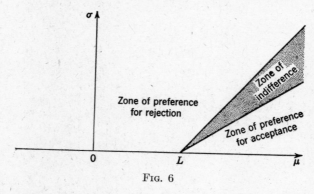

FIG. 6

preferred if the proportion of units in the lot for which $x \leq L$ exceeds .05, and no strong preference exists for either action if the proportion of units in the lot with $x \leq L$ lies between .01 and .05. The proportion of units with $x \leq L$ is greater than or equal to .05 if, and only if, $(\mu - L)/\sigma \leq \lambda_1$, and the proportion of such units is $\leq .01$ if, and only if, $(\mu - L)/\sigma \geq \lambda_2$ $(\lambda_1 < \lambda_2)$. The values λ_1 and λ_2 can be obtained from a table of the normal distribution. Thus the zone of preference for rejection is given by the set of all values μ and σ for which $(\mu - L)/\sigma \leqq \lambda_1$, the zone of preference for acceptance is given by $(\mu - L)/\sigma \geqq \lambda_2$, and the zone of indifference is given by $\lambda_1 < (\mu - L)/\sigma < \lambda_2$. These three zones are represented in Fig. 6, where μ is measured along the horizontal axis and σ along the vertical axis. The zone of indifference is bounded by two straight lines which go through the point L on the abscissa axis and have slopes $1/\lambda_1$ and $1/\lambda_2$, respectively.

2.3.2 Requirements Imposed on the OC Function

Suppose that the hypothesis H_0 to be tested states that the true parameter point θ lies in a given set ω of parameter points. Then we wish to make the probability of accepting H_0 as high as possible when θ lies in ω, and as low as possible when θ is outside ω. Since the probability of accepting H_0 is by definition equal to the OC function $L(\theta)$, an OC function is considered more desirable the higher the value of $L(\theta)$ for any θ in ω and the lower the value of $L(\theta)$ for any θ outside ω. An ideal OC function would be given by a function $L(\theta)$ such that $L(\theta) = 1$ for any θ in ω and $L(\theta) = 0$ for any θ outside ω. Suppose, for example, that there is only one unknown parameter θ and the hypothesis to be tested is the statement that $\theta \leqq \theta_0$. Then, an ideal OC function, as shown in Fig. 7, would be given by a function $L(\theta)$ such that $L(\theta) = 1$ for $\theta \leqq \theta_0$ and $L(\theta) = 0$ for $\theta > \theta_0$.

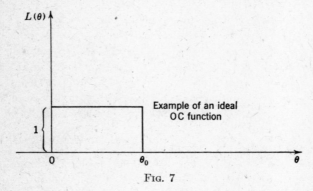

FIG. 7

The ideal form of the OC function can never be achieved on the basis of incomplete information about θ supplied by a random sample drawn from the population, but it can be approached arbitrarily closely if we are willing to take a sufficiently large sample.

The nearer the OC function is to the ideal function and the smaller the expected number of observations required, the more desirable is the sequential test. These two desirable features of a test are somewhat in conflict, since the closer we approach the ideal form of the OC function, the larger, in general, will be the number of observations required by the test. To achieve a compromise between these two conflicting desiderata, we may proceed as follows. First we formulate requirements concerning the closeness of the OC function to the ideal function and then consider only tests which satisfy these requirements. From these tests we try to select one for which the expected number of observations required by the test is as small as possible. To impose the desired conditions on the OC function first and then to minimize with respect to the expected number of observations does not seem to be an unreasonable procedure, since the OC function is perhaps of primary importance.

To formulate requirements on the OC function, we shall make use of the subdivision of the parameter space into the three zones discussed in the preceding section. Since in the zone of indifference there is no strong preference for one or the other action, we shall not impose any conditions on the behavior of $L(\theta)$ within the zone of indifference. In the zones of preference for acceptance and rejection the requirements on the OC function may reasonably be stated as follows. For any θ in the zone of preference for acceptance the probability of rejecting the hypothesis H_0, i.e., the value of $1 - L(\theta)$, should be less than or equal to a preassigned value α, and for any θ in the zone of preference for rejection the probability of accepting H_0, i.e., the value of $L(\theta)$, should be less than or equal to a preassigned value β.

We can summarize the requirements imposed on the OC function as follows. First the parameter space is subdivided into three mutually exclusive zones: a zone of preference for acceptance, a zone of preference for rejection, and a zone of indifference. Then two positive values α and β, both < 1, are selected. The requirements imposed on the OC function are then given by the two following conditions:

(2:1) $1 - L(\theta) \leqq \alpha$ for any θ in the zone of preference for acceptance

(2:2) $L(\theta) \leqq \beta$ for any θ in the zone of preference for rejection

Condition (2:1) can also be written as

(2:3) $L(\theta) \geqq 1 - \alpha$ for any θ in the zone of preference for acceptance

The subdivision of the parameter space into three zones, as well as the choice of the values α and β, is to be made on the basis of practical considerations in each particular case. We shall say that a sequential test is admissible if it satisfies the requirements (2:2) and (2:3).

A typical OC function satisfying the conditions (2:2) and (2:3) is shown in Fig. 8, where there is only one unknown parameter θ and the zone of preference for acceptance is defined by $\theta \leq \theta_0$, and the zone of preference for rejection is defined by $\theta \geq \theta_1$.   ($\theta_0 < \theta_1$.)

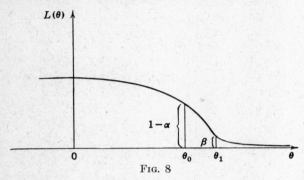

FIG. 8

2.3.3 The ASN Function as a Basis for the Selection of a Sequential Test

After the parameter space has been subdivided into three zones and the quantities α and β have been chosen, we consider only tests which are admissible, i.e., tests which satisfy the conditions (2:2) and (2:3). Clearly, we wish to select a sequential test for which the expected value of the number of observations required by the test is as small as possible. This expected value $E_\theta(n)$ depends, as we have seen in Section 2.2.2, on the parameter point θ. In section 2.2.2 we referred to the function $E_\theta(n)$ as the ASN function of the test.

The expected value $E_\theta(n)$ of the number of observations to be made depends, of course, also on the particular sequential test used. To put this dependence in evidence, we shall occasionally use the symbol $E_\theta(n \mid S)$ to denote the value $E_\theta(n)$ when the sequential test S is applied.

It is of particular interest to consider for any particular θ the minimum [2] value of $E_\theta(n \mid S)$ with respect to S where S may be any admissible sequential test. This minimum value, in symbols $\underset{S}{\text{Min}}\, E_\theta(n \mid S)$, depends only on θ. Clearly, for any admissible sequential test S' we have

$$E_\theta(n \mid S') \geq \underset{S}{\text{Min}}\, E_\theta(n \mid S)$$

If an admissible sequential test S_0 exists for which the expected value of the number of observations is minimized for all θ, i.e., for which

---

[2] If the minimum value does not exist, we can take the greatest lower bound with respect to S.

$E_\theta(n \mid S_0) = \underset{S}{\text{Min}} \, E_\theta(n \mid S)$ for *all* θ, then S_0 may be regarded as a "uniformly best" test. In general, however, no uniformly best test exists,[3] i.e., it will not be possible to minimize the expected value of the required number of observations simultaneously for *all* θ. Thus, in such cases some compromise principle is to be adopted for the selection of a sequential test. We do not propose to enter into a discussion of the various possible compromise principles that could be advanced, since the various possibilities have not yet been fully investigated. However, for the particular, but theoretically very interesting, case when a simple hypothesis is tested against a single alternative, the situation has been clarified and we shall discuss it in some detail in the next section.

2.4 The Case When a Simple Hypothesis H_0 Is Tested against a Single Alternative H_1

2.4.1 Efficiency of a Sequential Test

We shall consider only two values of the parameter θ, say θ_0 and θ_1. Let H_0 be the hypothesis that $\theta = \theta_0$ and let H_1 denote the hypothesis that $\theta = \theta_1$. We shall refer to H_0 as the null hypothesis and to H_1 as the alternative hypothesis. With any sequential test of the hypothesis H_0 against the alternative hypothesis H_1 there will be associated two numbers α and β between 0 and 1 such that if H_0 is true the probability is α that we shall commit an error of the first kind (we shall reject H_0), and if H_1 is true the probability is β that we shall commit an error of the second kind (we shall accept H_0). Two sequential tests S and S' will be said to be of equal strength if the values α and β associated with S are equal to the corresponding values α' and β' associated with S'. If $\alpha < \alpha'$ and $\beta \leq \beta'$, or if $\alpha \leq \alpha'$ and $\beta < \beta'$, we shall say that S is stronger than S' (S' is weaker than S). If $\alpha < \alpha'$ and $\beta > \beta'$, or if $\alpha > \alpha'$ and $\beta < \beta'$, we shall say that the strength of S is not comparable to that of S'.

Restricting ourselves to sequential tests of a given strength (α, β), a test may be regarded as more desirable the smaller the expected number of observations required by the test. If S and S' are two sequential tests of equal strength such that $E_{\theta_0}(n \mid S) \leq E_{\theta_0}(n \mid S')$ and $E_{\theta_1}(n \mid S) < E_{\theta_1}(n \mid S')$, or $E_{\theta_0}(n \mid S) < E_{\theta_0}(n \mid S')$ and $E_{\theta_1}(n \mid S) \leq E_{\theta_1}(n \mid S')$, the test S will be considered preferable to S'. If a test S_0 exists such that $E_{\theta_0}(n \mid S_0) \leq E_{\theta_0}(n \mid S)$ and $E_{\theta_1}(n \mid S_0) \leq E_{\theta_1}(n \mid S)$

---

[3] The situation here is similar to that in the Neyman-Pearson theory of testing hypotheses, where uniformly most powerful tests exist only in exceptional cases.

for all tests S of strength equal to that of S_0, we shall say that S_0 is an optimum test.

We shall denote by $n_0(\alpha, \beta)$ the minimum value of $E_{\theta_0}(n \mid S)$ with respect to S, and by $n_1(\alpha, \beta)$ the minimum value of $E_{\theta_1}(n \mid S)$ with respect to S, where S may be any sequential test of strength (α, β).[4] Then for any sequential test S of strength (α, β) we have $E_{\theta_0}(n \mid S) \geqq n_0(\alpha, \beta)$ and $E_{\theta_1}(n \mid S) \geqq n_1(\alpha, \beta)$. A sequential test S of strength (α, β) is an optimum test if $E_{\theta_0}(n \mid S) = n_0(\alpha, \beta)$ and $E_{\theta_1}(n \mid S) = n_1(\alpha, \beta)$. The existence of an optimum test has not been proved. However, it will be shown in Section A.7 of the Appendix that for the so-called sequential probability ratio test S_0 of strength (α, β), defined in Chapter 3, the ratios

$$(2{:}4) \qquad \frac{E_{\theta_0}(n \mid S_0)}{n_0(\alpha, \beta)} \quad \text{and} \quad \frac{E_{\theta_1}(n \mid S_0)}{n_1(\alpha, \beta)}$$

can exceed 1 only by very small quantities which can be neglected for practical purposes. Thus, for all practical purposes, the sequential probability ratio test may be regarded as an optimum test.[5] In Section A.7 it is also shown that the ratios (2:4) converge to 1 as θ_1 approaches θ_0.

We shall define the efficiency of a sequential test S of strength (α, β) by the ratio $\dfrac{n_0(\alpha, \beta)}{E_{\theta_0}(n \mid S)}$ when H_0 is true, and by $\dfrac{n_1(\alpha, \beta)}{E_{\theta_1}(n \mid S)}$ when H_1 is true. Clearly, the efficiency of a sequential test under H_0, as well as under H_1, lies always between 0 and 1. The greater the efficiency of a sequential test of a given strength the more desirable it is. An optimum test has the efficiency 1 under H_0, as well as under H_1. The sequential probability ratio test for testing H_0 against H_1 is shown in Section A.7 to have an efficiency, if not exactly, very nearly equal to 1 under H_0 as well as under H_1. As mentioned before, in Section A.7 it is shown that the efficiency of the sequential probability ratio test approaches 1 under H_0 as well as under H_1, when θ_1 approaches θ_0.

2.4.2 Efficiency of the Current Test Procedure, Viewed as a Particular Case of a Sequential Test

The current test procedure may be regarded as a particular case of a sequential test. In fact, if N denotes the fixed number of observations used in the current procedure and if W_N denotes the critical region,

---

[4] If the minimum value with respect to S does not exist, we take the greatest lower bound.

[5] The author conjectures that the sequential probability ratio test is exactly an optimum test, but he did not succeed in proving this.

i.e., W_N is the totality of all those samples of size N for which the hypothesis under test is rejected, then the current procedure may be regarded as a sequential test defined as follows. For all positive integral values $m < N$, the regions $R_m{}^0$, $R_m{}^1$ are the empty subsets of the m-dimensional sample space M_m, and $R_m = M_m$. For $m = N$, $R_N{}^1$ is equal to W_N, $R_N{}^0$ is equal to the totality of all samples of size N not contained in $R_N{}^1$, and R_N is the empty set. Thus, for the current test procedure we have $E_{\theta_0}(n) = E_{\theta_1}(n) = N$.

It will be shown later that the efficiency of the current test for testing H_0 against H_1, based on the most powerful critical region, is rather low. Frequently it is below $\frac{1}{2}$. In other words, an optimum sequential test can attain the same α and β as the current most powerful test on the basis of an expected number of observations much smaller than the fixed number of observations needed for the current most powerful test.

In Chapter 3 a simple sequential test procedure for testing H_0 against H_1 will be proposed. It is called the sequential probability ratio test, which for practical purposes can be regarded as an optimum sequential test. It will be seen that these sequential tests usually lead to average savings of about 50 per cent in the number of trials as compared with the current most powerful test.

Chapter 3. THE SEQUENTIAL PROBABILITY RATIO TEST FOR TESTING A SIMPLE HYPOTHESIS H_0 AGAINST A SINGLE ALTERNATIVE H_1

3.1 Definition of the Sequential Probability Ratio Test

Let $f(x, \theta)$ denote the distribution of the random variable x under consideration.[1] Let H_0 be the hypothesis that $\theta = \theta_0$, and H_1 the hypothesis that $\theta = \theta_1$. Thus, the distribution of x is given by $f(x, \theta_0)$ when H_0 is true, and by $f(x, \theta_1)$ when H_1 is true. We shall denote the successive observations on x by $x_1, x_2, \cdots$, etc.

As mentioned before, we consider only two cases: (1) x admits a probability density function; (2) x has a discrete distribution. It is our intention to cover both cases simultaneously. However, the difficulty arises that some statements will have to be formulated slightly differently, depending on whether x admits a density function or x has a discrete distribution. This difference in formulation is caused mostly by the fact that "probability density" in the continuous case is to be replaced by "probability" in the discrete case. For the sake of brevity, we shall occasionally use the word "probability" to mean "probability density" in the continuous case, if this can be done without danger of confusion. With this understanding it will frequently be possible to cover the discrete, as well as the continuous, case with a single statement.

For any positive integral value m the probability that a sample $x_1, \cdots, x_m$ is obtained is given by

$$p_{1m} = f(x_1, \theta_1) \cdots f(x_m, \theta_1)$$

when H_1 is true, and by

$$p_{0m} = f(x_1, \theta_0) \cdots f(x_m, \theta_0)$$

when H_0 is true.

The sequential probability ratio test for testing H_0 against H_1 is defined as follows: Two positive constants A and B ($B < A$) are chosen. At each stage of the experiment (at the mth trial for any integral value m), the probability ratio p_{1m}/p_{0m} is computed. If

(3:1) $$B < \frac{p_{1m}}{p_{0m}} < A$$

[1] $f(x, \theta)$ denotes the probability density function of x, if a density function exists. If x has a discrete distribution, $f(x, \theta)$ denotes the probability that the random variable under consideration takes the value x.

the experiment is continued by taking an additional observation. If

(3:2)
$$\frac{p_{1m}}{p_{0m}} \geqq A$$

the process is terminated with the rejection of H_0 (acceptance of H_1). If

(3:3)
$$\frac{p_{1m}}{p_{0m}} \geqq B$$

the process is terminated with the acceptance of H_0.[2]

The constants A and B are to be determined so that the test will have the prescribed strength (α, β). The relations among the quantities α, β, A, and B will be discussed in the next section.

For purposes of practical computation, it is much more convenient to compute the logarithm of the ratio p_{1m}/p_{0m} than the ratio p_{1m}/p_{0m} itself. The reason for this is that $\log (p_{1m}/p_{0m})$ can be written as the sum of m terms, i.e.,

(3:4)
$$\log \frac{p_{1m}}{p_{0m}} = \log \frac{f(x_1, \theta_1)}{f(x_1, \theta_0)} + \cdots + \log \frac{f(x_m, \theta_1)}{f(x_m, \theta_0)}$$

We shall denote the ith term in this sum by z_i, i.e.,

(3:5)
$$z_i = \log \frac{f(x_i, \theta_1)}{f(x_i, \theta_0)}$$

The test procedure is carried out as follows, the quantities z_i $(i = 1, 2, \cdots)$ being used: At each stage of the experiment (at the mth trial for each integral value of m), the cumulative sum $z_1 + \cdots + z_m$ is computed. If

(3:6)
$$\log B < z_1 + \cdots + z_m < \log A$$

the experiment is continued by taking an additional observation. If

$$z_1 + \cdots + z_m \geqq \log A$$

the process is terminated with the rejection of H_0. If

$$z_1 + \cdots + z_m \leqq \log B$$

the process is terminated with the acceptance of H_0.

[2] If for a particular sample $p_{1m} = p_{0m} = 0$, we shall define the value of the ratio p_{1m}/p_{0m} as 1. If for some sample $(x_1, \cdots, x_m)$ we have $p_{1m} > 0$ but $p_{0m} = 0$, inequality (3:2) is considered fulfilled and H_0 is rejected.

A few simple illustrations will help to make the procedure more concrete. Suppose that the random variable x can take only two values, 0 and 1. We shall denote the probability that $x = 1$ by p, the value of which is assumed to be unknown. Thus, p is the unknown parameter of the distribution. The distribution of x is given by the function $f(x, p)$ which is defined only for two values of x, namely $x = 0$ and $x = 1$. $f(1, p) = p$ and $f(0, p) = 1 - p$. Let H_0 be the hypothesis that $p = p_0$ and H_1 the hypothesis that $p = p_1$ $(p_1 \neq p_0)$. Then

$$z_i = \log \frac{f(x_i, p_1)}{f(x_i, p_0)} = \log \frac{p_1}{p_0} \text{ if } x_i = 1$$

$$= \log \frac{1 - p_1}{1 - p_0} \text{ if } x_i = 0$$

Hence,

$$(3{:}7) \qquad z_1 + \cdots + z_m = m^* \log \frac{p_1}{p_0} + (m - m^*) \log \frac{1 - p_1}{1 - p_0}$$

where m^* denotes the number of ones in the sequence of the first m observations. We accept H_0 if

$$m^* \log \frac{p_1}{p_0} + (m - m^*) \log \frac{1 - p_1}{1 - p_0} \leqq \log B$$

We reject H_0 (accept H_1) if

$$m^* \log \frac{p_1}{p_0} + (m - m^*) \log \frac{1 - p_1}{1 - p_0} \geqq \log A$$

We continue the experiment by taking an additional observation if

$$\log B < m^* \log \frac{p_1}{p_0} + (m - m^*) \log \frac{1 - p_1}{1 - p_0} < \log A$$

The expression (3:7) can, of course, be obtained cumulatively. If an observation is a one, the constant $\log (p_1/p_0)$ is added to the preceding value of (3:7) to obtain the new value. If the observation is a zero, the constant $\log (1 - p_1)/(1 - p_0)$ is added.

As a second example, consider the problem of testing a hypothesis about the mean of a normal distribution. Let x be a normally distributed random variable with unknown mean θ and unit variance.

Let H_0 be the hypothesis that $\theta = \theta_0$ and H_1 the hypothesis that $\theta = \theta_1$. Then

$$f(x, \theta_0) = \frac{1}{\sqrt{2\pi}} e^{-\frac{1}{2}(x-\theta_0)^2}$$

and

$$f(x, \theta_1) = \frac{1}{\sqrt{2\pi}} e^{-\frac{1}{2}(x-\theta_1)^2}$$

Hence,

$$z_i = \log \frac{f(x_i, \theta_1)}{f(x_i, \theta_0)} = (\theta_1 - \theta_0)x_i + \frac{1}{2}(\theta_0{}^2 - \theta_1{}^2)$$

and

$$\log \frac{p_{1m}}{p_{0m}} = z_1 + \cdots + z_m = (\theta_1 - \theta_0) \sum_{i=1}^{m} x_i + \frac{m}{2}(\theta_0{}^2 - \theta_1{}^2)$$

If

$$(\theta_1 - \theta_0) \sum_{1}^{m} x_i + \frac{m}{2}(\theta_0{}^2 - \theta_1{}^2) \geqq \log A$$

the process is terminated with the rejection of H_0.   If

$$(\theta_1 - \theta_0) \sum_{1}^{m} x_i + \frac{m}{2}(\theta_0{}^2 - \theta_1{}^2) \leqq \log B$$

the process is terminated with the acceptance of H_0.   If

$$\log B < (\theta_1 - \theta_0) \sum_{1}^{m} x_i + \frac{m}{2}(\theta_0{}^2 - \theta_1{}^2) < \log A$$

the experiment is continued by taking an additional observation. Again, $\log (p_{1m}/p_{0m})$ can be computed cumulatively if after each observation x_i we compute $(\theta_1 - \theta_0)x_i + \frac{1}{2}(\theta_0{}^2 - \theta_1{}^2)$ and add it to the preceding value of $\log (p_{1m}/p_{0m})$.

3.2   Fundamental Relations among the Quantities α, β, A, and B

In this section we shall derive certain inequalities satisfied by the quantities α, β, A, and B which will provide the basis for determining the constants A and B in the sequential probability ratio test.

We shall say a sample $(x_1, \cdots, x_n)$ is of type 0 if

$$B < \frac{p_{1m}}{p_{0m}} = \frac{f(x_1, \theta_1) \cdots f(x_m, \theta_1)}{f(x_1, \theta_0) \cdots f(x_m, \theta_0)} < A \text{ for } m = 1, \cdots, n-1$$

and

$$\frac{p_{1n}}{p_{0n}} \leqq B$$

Similarly, we shall say a sample $(x_1, \cdots, x_n)$ is of type 1 if

$$B < \frac{p_{1m}}{p_{0m}} = \frac{f(x_1, \theta_1) \cdots f(x_m, \theta_1)}{f(x_1, \theta_0) \cdots f(x_m, \theta_0)} < A \text{ for } m = 1, \cdots, n-1$$

and

$$\frac{p_{1n}}{p_{0n}} \geqq A$$

Thus, a sample of type 0 leads to the acceptance of H_0 and a sample of type 1 leads to the acceptance of H_1 (rejection of H_0).

Clearly, for any given sample $(x_1, \cdots, x_n)$ of type 1 the probability of obtaining such a sample is at least A times as large under hypothesis H_1 as under hypothesis H_0. Thus, the probability measure of the totality of all samples of type 1 is also at least A times as large under H_1 as under H_0. The probability measure of the totality of all samples of type 1 is the same as the probability that the sequential process will terminate with the acceptance of H_1 (rejection of H_0). But the latter probability is equal to α when H_0 is true and to $1 - \beta$ when H_1 is true.[3] Thus, we obtain the inequality

$$(3:8) \qquad\qquad 1 - \beta \geqq A\alpha$$

This inequality can be written as

$$(3:9) \qquad\qquad A \leqq \frac{1 - \beta}{\alpha}$$

Thus, $(1 - \beta)/\alpha$ is an upper limit for A.

A lower limit for B can be derived in a similar way. In fact, for any given sample $(x_1, \cdots, x_n)$ of type 0 the probability of obtaining such a sample under H_1 is at most B times as large as the probability of obtaining such a sample when H_0 is true. Thus, also the probability of accepting H_0 is at most B times as large when H_1 is true as when H_0 is true. Since the probability of accepting H_0 is $1 - \alpha$ when H_0 is true and β when H_1 is true, we obtain the inequality

$$(3:10) \qquad\qquad \beta \leqq (1 - \alpha)B$$

This inequality can be written as

$$(3:11) \qquad\qquad B \geqq \frac{\beta}{1 - \alpha}$$

Thus, $\beta/(1 - \alpha)$ is a lower limit for B.

---

[3] The probability that H_0 will be accepted when H_1 is true is by definition equal to β. Section A.1 of the Appendix shows that the probability is one that the sequential process will eventually terminate. Thus, the probability that H_0 will be rejected when H_1 is true must be equal to $1 - \beta$.

Inequalities (3:8) and (3:10) can also be written as

(3:12)
$$\frac{\alpha}{1 - \beta} \leqq \frac{1}{A}$$

and

(3:13)
$$\frac{\beta}{1 - \alpha} \leqq B$$

These inequalities are of considerable value in practical applications, since they furnish upper limits for α and β for given values of A and B. For example, it follows from these inequalities that

(3:14)
$$\alpha \leqq \frac{1}{A}$$

and

(3:15)
$$\beta \leqq B$$

It may be of interest to represent graphically the totality of all pairs (α, β) which satisfy the inequalities (3:12) and (3:13). Any pair

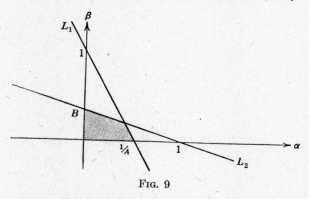

FIG. 9

(α, β) can be represented by a point in the plane with abscissa α and ordinate β. Consider the straight lines L_1 and L_2 in the plane given by the equations

(3:16)
$$\alpha A = 1 - \beta$$

and

(3:17)
$$\beta = B(1 - \alpha)$$

respectively. The line L_1 intersects the abscissa axis at $\alpha = (1/A)$ and the ordinate axis at $\beta = 1$. Similarly, the line L_2 intersects the abscissa axis at $\alpha = 1$ and the ordinate axis at $\beta = B$. The region consisting of all points (α, β) which satisfy the inequalities (3:12) and (3:13) is the interior and the boundary of the quadrilateral determined by the lines L_1, L_2, and the coordinate axes. This region is shown by the shaded area in Fig. 9.

The inequalities (3:12) and (3:13) have been derived under the assumption that the successive observations $x_1, x_2, \cdots$, etc., are independent observations on x. The assumption of the independence of the observations has been used in showing that the probability is one that the sequential process will eventually terminate.[4] The rest of the derivation, however, remains valid also when the successive observations are dependent, i.e., when the conditional distribution of the ith observation x_i is affected by the outcome of the preceding observations $x_1, \cdots, x_{i-1}$. If the successive observations are not independent, the probability that a sample $(x_1, \cdots, x_m)$ will be obtained, i.e., the joint distribution of $(x_1, \cdots, x_m)$, is no longer given by the product $f(x_1, \theta)f(x_2, \theta) \cdots f(x_m, \theta)$, but by a more general function $p_m(x_1, \cdots, x_m)$. Thus, in dealing with dependent observations, the null hypothesis H_0 will be the statement that the distribution of the sample $(x_1, \cdots, x_m)$ is given by some function $p_{0m}(x_1, \cdots, x_m)$, and the alternative hypothesis H_1 will be the statement that this distribution is given by some other function $p_{1m}(x_1, \cdots, x_m)$. We can construct the sequential probability ratio test for testing H_0 against H_1 in the same way as for independent observations. That is to say, we select two constants A and B $(B < A)$ and continue taking observations as long as $B < \dfrac{p_{1m}(x_1, \cdots, x_m)}{p_{0m}(x_1, \cdots, x_m)} < A$. The first time that the probability ratio $p_{1m}/p_{0m} \geqq A$ or $\leqq B$, we terminate the sequential process. H_0 is accepted if $p_{1m}/p_{0m} \leqq B$ and H_1 is accepted if $p_{1m}/p_{0m} \geqq A$. The fundamental inequalities (3:12) and (3:13) remain valid for such a test procedure in spite of the dependence of the successive observations, provided that the probability is one that the procedure will eventually terminate. It can be shown that for a very general class of joint distributions $p_{0m}(x_1, \cdots, x_m)$ and $p_{1m}(x_1, \cdots, x_m)$ the probability is one that the procedure will eventually terminate. Thus, the validity of the inequalities (3:12) and (3:13) is by no means restricted to the case of independent observations. They are generally valid also for dependent observations.

A simple case of dependent observations arises when we sample from a finite population. Suppose, for example, that a lot consisting of N units of a manufactured product is submitted for acceptance inspection.

[4] See Section A.1 in the Appendix.

Let D be the number of defectives in the lot, which is assumed to be unknown. To each defective unit we assign the value 1 and to each non-defective unit the value 0. Then the distribution of a single observation x is given by $f(x, p)$ where $f(1, p) = p$, $f(0, p) = 1 - p$, and $p = D/N$. The successive observations are, however, not independent. For example, if $x_1 = 1$, the distribution of x_2 is given by $f\left(x_2, \dfrac{D-1}{N-1}\right)$, while if $x_1 = 0$, the distribution of x_2 is given by $f\left(x_2, \dfrac{D}{N-1}\right)$. If we denote by d_i the number of defectives (the number of ones) in the set of the first i observations $x_1, \cdots, x_i$, the joint distribution of $(x_1, \cdots, x_m)$ is given by [5]

(3:18)
$$p_m = f\left(x_1, \frac{D}{N}\right) f\left(x_2, \frac{D-d_1}{N-1}\right) f\left(x_3, \frac{D-d_2}{N-2}\right) \cdots f\left(x_m, \frac{D-d_{m-1}}{N-m+1}\right)$$

Suppose that the hypothesis H_0 is that D is equal to some specified value D_0, and H_1 is the hypothesis that D is equal to some value D_1 $(D_1 > D_0)$. Then the distribution of $(x_1, \cdots, x_m)$ under H_0 is given by

(3:19) $$p_{0m} = f\left(x_1, \frac{D_0}{N}\right) f\left(x_2, \frac{D_0-d_1}{N-1}\right) \cdots f\left(x_m, \frac{D_0-d_{m-1}}{N-m+1}\right)$$

and the distribution under H_1 by

(3:20) $$p_{1m} = f\left(x_1, \frac{D_1}{N}\right) f\left(x_2, \frac{D_1-d_1}{N-1}\right) \cdots f\left(x_m, \frac{D_1-d_{m-1}}{N-m+1}\right)$$

The sequential probability ratio test for testing H_0 against H_1 is based on the ratio p_{1m}/p_{0m}. Inspection continues as long as $B < p_{1m}/p_{0m} < A$. The lot is accepted if $p_{1m}/p_{0m} \leqq B$ and the lot is rejected if $p_{1m}/p_{0m} \geqq A$. The fundamental inequalities (3:12) and (3:13) remain valid for this test procedure in spite of the dependence of the observations.

3.3 Determination of the Constants A and B in Practice

Suppose that we wish to have a test procedure of strength (α, β). Then our problem is to determine the constants A and B such that the resulting test will have the desired strength (α, β). Let us denote by $A(\alpha, \beta)$ and $B(\alpha, \beta)$ the values of A and B, respectively, for which the test has the required strength (α, β). The exact determination of

---

[5] This formula is valid as long as $d_{m-1} \leqq D$.

the values $A(\alpha, \beta)$ and $B(\alpha, \beta)$ is usually very laborious.[6] However, the fundamental inequalities derived in the preceding section permit an approximate determination of A and B which will suffice for most practical purposes. From (3:9) and (3:11) it follows that

$$(3:21) \qquad\qquad A(\alpha, \beta) \leqq \frac{1 - \beta}{\alpha}$$

and

$$(3:22) \qquad\qquad B(\alpha, \beta) \geqq \frac{\beta}{1 - \alpha}$$

We shall propose to put $A = (1 - \beta)/\alpha = a(\alpha, \beta)$, say, and $B = \beta/(1 - \alpha) = b(\alpha, \beta)$, say, and we shall investigate the consequences of this determination of A and B. From (3:21) and (3:22) it follows that the value $a(\alpha, \beta)$ chosen for A is greater than or equal to the exact value $A(\alpha, \beta)$, and the value $b(\alpha, \beta)$ chosen for B is less than or equal to the exact value $B(\alpha, \beta)$. Then, letting $A = a(\alpha, \beta)$ instead of $A(\alpha, \beta)$ and $B = b(\alpha, \beta)$ instead of $B(\alpha, \beta)$ will, in general, change the probabilities of errors of the first and second kinds. If A were put equal to a value greater than $A(\alpha, \beta)$, and if B were put equal to $B(\alpha, \beta)$, then the resulting probability of an error of the first kind would be less than α, but the probability of an error of the second kind would be slightly larger than β. Similarly, if we were to use the exact value $A(\alpha, \beta)$ for A, but a value B below the exact value $B(\alpha, \beta)$, the resulting probability of an error of the second kind would be less than β, and the probability of an error of the first kind would be slightly greater than α. Thus, if a value A is used which is higher than the exact value $A(\alpha, \beta)$ and a value B is used which is lower than the exact value $B(\alpha, \beta)$, it is not clear what the resulting effect on the probabilities of errors of the first and second kinds will be. Let us denote by α' and β' the resulting probabilities of errors of the first and second kinds when $A = a(\alpha, \beta)$ and $B = b(\alpha, \beta)$. From (3:12) and (3:13) it follows that

$$(3:23) \qquad\qquad \frac{\alpha'}{1 - \beta'} \leqq \frac{1}{a(\alpha, \beta)} = \frac{\alpha}{1 - \beta}$$

and

$$(3:24) \qquad\qquad \frac{\beta'}{1 - \alpha'} \leqq b(\alpha, \beta) = \frac{\beta}{1 - \alpha}$$

---

[6] The results in Section A.4 of the Appendix can be used for deriving arbitrarily close approximations to the values $A(\alpha, \beta)$ and $B(\alpha, \beta)$.

From these inequalities it follows that

(3:25)
$$\alpha' \leq \frac{\alpha}{1 - \beta}$$

and

(3:26)
$$\beta' \leq \frac{\beta}{(1 - \alpha)}$$

Multiplying (3:23) by $(1 - \beta)(1 - \beta')$ and (3:24) by $(1 - \alpha)(1 - \alpha')$ and adding the two resulting inequalities, we obtain

(3:27)
$$\alpha' + \beta' \leq \alpha + \beta$$

Inequalities (3:25), (3:26), and (3:27) give valuable upper limits for α' and β'. The values α and β will usually be small in practical applications. Most frequently they will lie in the range from .01 to .05. Thus, $\alpha/(1 - \beta)$ and $\beta/(1 - \alpha)$ will be very nearly equal to α and β, respectively. It follows then from (3:25) and (3:26) that the amount by which α' may exceed α, or β' may exceed β is very small and can be neglected for all practical purposes. Moreover, (3:27) shows that at least one of the inequalities $\alpha' \leq \alpha$ and $\beta' \leq \beta$ must hold exactly. In other words, by using $a(\alpha, \beta)$ and $b(\alpha, \beta)$ instead of $A(\alpha, \beta)$ and $B(\alpha, \beta)$, respectively, at most one of the probabilities α and β may be increased.

Thus, we may conclude: *The use of $a(\alpha, \beta)$ and $b(\alpha, \beta)$ instead of $A(\alpha, \beta)$ and $B(\alpha, \beta)$, respectively, cannot result in any appreciable increase in the value of either α or β. In other words, for all practical purposes the test corresponding to $A = a(\alpha, \beta)$ and $B = b(\alpha, \beta)$ provides at least the same protection against wrong decisions as the test corresponding to $A = A(\alpha, \beta)$ and $B = B(\alpha, \beta)$.*

Our discussion so far leaves still open the possibility that the use of $a(\alpha, \beta)$ and $b(\alpha, \beta)$ instead of $A(\alpha, \beta)$ and $B(\alpha, \beta)$, respectively, may result in an appreciable decrease of α, or β, or both. If this were so, it would mean only that the test corresponding to $A = a(\alpha, \beta)$ and $B = b(\alpha, \beta)$ would provide a better protection against wrong decisions than the test corresponding to $A = A(\alpha, \beta)$ and $B = B(\alpha, \beta)$. Thus, the only disadvantage that may arise from using $a(\alpha, \beta)$ and $b(\alpha, \beta)$ instead of $A(\alpha, \beta)$ and $B(\alpha, \beta)$, respectively, is that it may result in an appreciable increase in the number of observations required by the test. In fact, since $a(\alpha, \beta) \geq A(\alpha, \beta)$ and $b(\alpha, \beta) \leq B(\alpha, \beta)$, the number of observations required by the test corresponding to $A = a(\alpha, \beta)$ and $B = b(\alpha, \beta)$ can never be smaller than the number of observations required by the test corresponding to $A = A(\alpha, \beta)$ and $B = B(\alpha, \beta)$. Thus, if the increase in the necessary number of observations caused by the use of $a(\alpha, \beta)$ and $b(\alpha, \beta)$ instead of $A(\alpha, \beta)$ and $B(\alpha, \beta)$ can be shown to be only slight and of no practical consequence, the test corresponding to $A = a(\alpha, \beta)$ and $B = b(\alpha, \beta)$ serves the purpose just as well, and the determination of the exact values $A(\alpha, \beta)$ and $B(\alpha, \beta)$ is of little interest.

We shall now indicate the reasons why the increase in the necessary number of observations caused by the use of $a(\alpha, \beta)$ and $b(\alpha, \beta)$ instead of the exact values $A(\alpha, \beta)$ and $B(\alpha, \beta)$ will generally be only slight.[7] The reason that (3:21) and (3:22) are inequalities instead of equalities is that the sequential process may terminate with $p_{1m}/p_{0m} > A$ or $p_{1m}/p_{0m} < B$. If at the final stage p_{1m}/p_{0m} were exactly equal to A or B, then $A(\alpha, \beta)$ and $B(\alpha, \beta)$ would be exactly equal to $(1 - \beta)/\alpha$ and $\beta/(1 - \alpha)$, respectively. On the other hand, a possible excess of p_{1m}/p_{0m} over the boundaries A and B at the termination of the test procedure is caused only by the discontinuity of the number of observations, i.e., by the fact that the number of observations can take only integral values. Thus, if fractional observations were possible, i.e., if the number m of observations were a continuous variable, p_{1m}/p_{0m} would also be a continuous function of m and consequently $A(\alpha, \beta)$ and $B(\alpha, \beta)$ would be exactly equal to $a(\alpha, \beta)$ and $b(\alpha, \beta)$, respectively. That the increase in the necessary number of trials caused by the use of $a(\alpha, \beta)$ and $b(\alpha, \beta)$ will generally be slight is strongly indicated by the fact that the discrepancy between $A(\alpha, \beta)$ and $a(\alpha, \beta)$, as well as that between $B(\alpha, \beta)$ and $b(\alpha, \beta)$, arises only from the discontinuity of the number of observations. In Section 3.9 we give upper estimates of the increase in the expected number of trials caused by the use of $a(\alpha, \beta)$ and $b(\alpha, \beta)$. Numerical computations given in that section show that the increase is slight. It may be added that the nearer the distribution $f(x, \theta_1)$ is to the distribution $f(x, \theta_0)$ the smaller will be this increase in the expected number of trials. The reason for this is that the nearer $f(x, \theta_1)$ is to $f(x, \theta_0)$, the smaller the expected excess of p_{1m}/p_{0m} over the boundaries A and B and, therefore, also the smaller the discrepancy between $A(\alpha, \beta)$ and $a(\alpha, \beta)$ as well as that between $B(\alpha, \beta)$ and $b(\alpha, \beta)$. If $f(x, \theta_1)$ approaches $f(x, \theta_0)$ the exact values $A(\alpha, \beta)$ and $B(\alpha, \beta)$ converge to $a(\alpha, \beta)$ and $b(\alpha, \beta)$, respectively.

Hence, if experimentation is not excessively costly, for all practical purposes the following procedure may be adopted: *If a sequential test is desired such that the probability of an error of the first kind does not exceed α, and the probability of an error of the second kind does not exceed β, put $A = (1 - \beta)/\alpha$ and $B = \beta/(1 - \alpha)$ and carry out the sequential probability ratio test as defined by the inequalities (3:1), (3:2), and (3:3).*

---

[7] For a more complete discussion see Section 3.9.

The fact that for practical purposes we may put $A = a(\alpha, \beta)$ and $B = b(\alpha, \beta)$ brings out a surprising feature of the sequential test as compared with current tests. Whereas current tests cannot be carried out without finding the probability distribution of the statistic on which the test is based, there are no distribution problems in carrying out a sequential test. In fact, $a(\alpha, \beta)$ and $b(\alpha, \beta)$ depend on α and β only, and the ratio p_{1m}/p_{0m} can be calculated from the data of the problem without solving any distribution problems. Distribution problems arise in connection with the sequential process only if it is desired to find the probability distribution of the number of trials necessary for reaching a final decision. But this is of secondary importance as long as we know that the sequential test on the average leads to a saving in the number of observations.

3.4 The OC Function of the Sequential Probability Ratio Test [8]

Since the sequential probability ratio test for testing the hypothesis H_0 against the hypothesis H_1 will be applied to problems when the parameter θ can take values $\neq \theta_0$ and $\neq \theta_1$, it is of interest to derive the whole operating characteristic function $L(\theta)$ of the test. For convenience, we shall treat the case of a single unknown parameter θ in this section and in Section 3.5. The results can be extended without difficulty to any number of parameters. In Section 2.2.1, $L(\theta)$ has been defined as the probability that the sequential process will terminate with the acceptance of H_0 when θ is the true value of the parameter. In this section we shall indicate the derivation of an approximation formula for $L(\theta)$, neglecting the excess of p_{1m}/p_{0m} over the boundaries A and B at the termination of the process. A rigorous derivation (using a different method) together with upper and lower limits for the OC function will be given in Section A.2.3 of the Appendix.

Consider the expression

$$(3:28) \qquad \left[\frac{f(x, \theta_1)}{f(x, \theta_0)}\right]^{h(\theta)}$$

For each value θ, the value of $h(\theta)$ is determined so that $h(\theta) \neq 0$ and the expected value of the expression (3:28) is equal to 1, i.e.,

$$(3:29a) \qquad \int_{-\infty}^{+\infty} \left[\frac{f(x, \theta_1)}{f(x, \theta_0)}\right]^{h(\theta)} f(x, \theta) \, dx = 1$$

---

[8] As mentioned in the Introduction, the operating characteristic function for the special case of a binomial distribution was found by Milton Friedman and George W. Brown independently of each other, and slightly earlier by C. M. Stockman in England. The derivation of the OC function in the general case is due to the author.

if $f(x, \theta)$ is the probability density function, or

(3:29b)
$$\sum_x \left[\frac{f(x, \theta_1)}{f(x, \theta_0)}\right]^{h(\theta)} f(x, \theta) = 1$$

if x has a discrete distribution (the summation is taken over all possible values of x).   It is shown in Section A.2.1 of the Appendix that under some slight restriction on the nature of the distribution function $f(x, \theta)$, there exists exactly one value $h(\theta) \neq 0$ such that (3:29) is fulfilled.

Hence, for any given value θ, the function of x given by

(3:30)
$$f^*(x, \theta) = \left[\frac{f(x, \theta_1)}{f(x, \theta_0)}\right]^{h(\theta)} f(x, \theta)$$

is a distribution function.

Since $h(\theta) \neq 0$, there are two possibilities: $h(\theta) > 0$ or $h(\theta) < 0$.   We shall first consider the case when $h(\theta) > 0$.

Let H denote the hypothesis that $f(x, \theta)$ is the true distribution of x and H^* the hypothesis that $f^*(x, \theta)$ is the true distribution of x. Consider the sequential probability ratio test S^* for testing H against H^* defined as follows: Continue taking observations as long as

(3:31)
$$B^{h(\theta)} < \frac{f^*(x_1, \theta) \cdots f^*(x_m, \theta)}{f(x_1, \theta) \cdots f(x_m, \theta)} < A^{h(\theta)}$$

Accept the hypothesis H if

(3:32)
$$\frac{f^*(x_1, \theta) \cdots f^*(x_m, \theta)}{f(x_1, \theta) \cdots f(x_m, \theta)} \leqq B^{h(\theta)}$$

Reject the hypothesis H (accept H^*) if

(3:33)
$$\frac{f^*(x_1, \theta) \cdots f^*(x_m, \theta)}{f(x_1, \theta) \cdots f(x_m, \theta)} \geqq A^{h(\theta)}$$

Since

(3:34)
$$\frac{f^*(x, \theta)}{f(x, \theta)} = \left[\frac{f(x, \theta_1)}{f(x, \theta_0)}\right]^{h(\theta)}$$

and since $h(\theta) > 0$, the inequalities (3:31), (3:32), and (3:33) are equivalent to

(3:35)
$$B < \frac{f(x_1, \theta_1) \cdots f(x_m, \theta_1)}{f(x_1, \theta_0) \cdots f(x_m, \theta_0)} < A$$

(3:36)
$$\frac{f(x_1, \theta_1) \cdots f(x_m, \theta_1)}{f(x_1, \theta_0) \cdots f(x_m, \theta_0)} \leqq B$$

and

(3:37)
$$\frac{f(x_1, \theta_1) \cdots f(x_m, \theta_1)}{f(x_1, \theta_0) \cdots f(x_m, \theta_0)} \geqq A$$

But these inequalities are identical with those defining the sequential probability ratio test S for testing H_0 against H_1, when the constants A and B are used. Thus, if the test S^* leads to the acceptance of H, the test S leads to the acceptance of H_0, and if S^* leads to the rejection of H, then S also leads to the rejection of H_0. From this, it follows that the probability of accepting H_0 when θ is true, i.e., the value of $L(\theta)$, is the same as the probability that the test S^* will lead to the acceptance of H when $f(x, \theta)$ is the true distribution of x.

To calculate the latter probability we shall apply the formulas (3:9) and (3:11) to the test procedure S^*. Denote by α' the probability that S^* will lead to the rejection of H when H is true, and by β' the probability that S^* leads to the acceptance of H when H^* is true. Applying the formulas (3:9) and (3:11) to the test procedure S^* we obtain

$$(3:38) \qquad A^{h(\theta)} \leqq \frac{1 - \beta'}{\alpha'}$$

and

$$(3:39) \qquad B^{h(\theta)} \geqq \frac{\beta'}{1 - \alpha'}$$

When the excess over the boundaries at the termination of the process is neglected, the equality sign holds in (3:38) and (3:39), that is,[9]

$$(3:40) \qquad A^{h(\theta)} \sim \frac{1 - \beta'}{\alpha'}$$

and

$$(3:41) \qquad B^{h(\theta)} \sim \frac{\beta'}{1 - \alpha'}$$

From (3:40) and (3:41) we obtain

$$(3:42) \qquad \alpha' \sim \frac{1 - B^{h(\theta)}}{A^{h(\theta)} - B^{h(\theta)}}$$

Since $\alpha' = 1 - L(\theta)$, we get

$$(3:43) \qquad L(\theta) \sim \frac{A^{h(\theta)} - 1}{A^{h(\theta)} - B^{h(\theta)}}$$

The case $h(\theta) < 0$ can be treated in a similar way. We obtain the same result, i.e., the approximation formula (3:43) remains valid also when $h(\theta) < 0$.

[9] The symbol $\sim$ indicates an approximate equality.

It is interesting to note that $h(\theta_0) = 1$ and $h(\theta_1) = -1$. This follows easily from (3:29b).

As an illustration, we shall determine $L(\theta)$ for the binomial case when x can take only the values 0 and 1 and the distribution $f(x, \theta)$ is given as follows: $f(1, \theta) = \theta$ and $f(0, \theta) = 1 - \theta$. Then equation (3:29b) can be written as

$$(3:44) \qquad \theta \left(\frac{\theta_1}{\theta_0}\right)^{h(\theta)} + (1 - \theta) \left(\frac{1 - \theta_1}{1 - \theta_0}\right)^{h(\theta)} = 1$$

To plot the OC function, it is not necessary to solve equation (3:44) with respect to $h(\theta)$. We may consider $h = h(\theta)$ a parameter and solve (3:44) with respect to θ. Then we obtain

$$(3:45) \qquad \theta = \frac{1 - \left(\dfrac{1 - \theta_1}{1 - \theta_0}\right)^{h}}{\left(\dfrac{\theta_1}{\theta_0}\right)^{h} - \left(\dfrac{1 - \theta_1}{1 - \theta_0}\right)^{h}}$$

If we let $A = (1 - \beta)/\alpha$ and $B = \beta/(1 - \alpha)$, (3:43) can be written as

$$(3:46) \qquad L(\theta) \sim \frac{\left(\dfrac{1 - \beta}{\alpha}\right)^{h} - 1}{\left(\dfrac{1 - \beta}{\alpha}\right)^{h} - \left(\dfrac{\beta}{1 - \alpha}\right)^{h}}$$

For any arbitrarily chosen value h, the point $[\theta, L(\theta)]$, computed from (3:45) and (3:46), will be a point on the OC function. The OC function can be drawn by plotting a sufficiently large number of points $[\theta, L(\theta)]$ corresponding to various values of h.

A typical OC function for the binomial case is shown in Fig. 10.

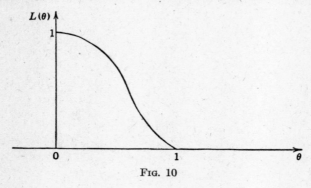

FIG. 10

We shall now compute $L(\theta)$ when x is normally distributed with unknown mean θ and known variance σ^2. In this case we have

$$f(x, \theta) = \frac{1}{\sqrt{2\pi}\sigma} e^{-\frac{1}{2\sigma^2}(x-\theta)^2}$$

The quantity $h(\theta)$ is the non-zero root of the equation

$$(3:47) \qquad \int_{-\infty}^{+\infty} \frac{1}{\sqrt{2\pi}\sigma} e^{-\frac{1}{2\sigma^2}(x-\theta)^2} \left[\frac{e^{-\frac{1}{2\sigma^2}(x-\theta_1)^2}}{e^{-\frac{1}{2\sigma^2}(x-\theta_0)^2}} \right]^{h(\theta)} dx = 1$$

Evaluating the above integral and solving the equation with respect to $h(\theta)$, we obtain

$$(3:48) \qquad h(\theta) = \frac{\theta_1 + \theta_0 - 2\theta}{\theta_1 - \theta_0}$$

An approximation to the OC function is obtained from (3:43) by substituting $(\theta_1 + \theta_0 - 2\theta)/(\theta_1 - \theta_0)$ for $h(\theta)$.

3.5   The ASN Function of a Sequential Probability Ratio Test

Let n denote the number of observations required by the test and let $E_\theta(n)$ be the expected value of n when θ is the true value of the parameter. This expected value $E_\theta(n)$ is a function of θ which we have called the average sample number function, or briefly the ASN function. In this section we shall outline the derivation of an approximation formula for the ASN function, neglecting the excess of p_{1m}/p_{0m} over the boundaries A and B at the termination of the sequential process. A more complete discussion together with upper and lower limits for the ASN function is given in Section A.3 of the Appendix.

Let N be an integer sufficiently large to allow the probability that $n \geq N$ to be neglected.[10] Thus we shall assume that $n < N$. Then we can write

$$(3:49) \quad z_1 + \cdots + z_N = (z_1 + \cdots + z_n) + (z_{n+1} + \cdots + z_N)$$

where

$$(3:50) \qquad z_\alpha = \log \frac{f(x_\alpha, \theta_1)}{f(x_\alpha, \theta_0)}$$

[10] It is shown in Section A.3.1 that no error is involved in assuming this, since we pass to the limit when N approaches ∞.

Taking expected values on both sides of (3:49), we obtain

$$(3:51) \qquad NE(z) = E(z_1 + \cdots + z_n) + E(z_{n+1} + \cdots + z_N)$$

where

$$(3:52) \qquad z = \log \frac{f(x, \theta_1)}{f(x, \theta_0)}$$

Since, for $\alpha > n$, the random variable z_α is distributed independently of n, the expected value of $z_{n+1} + \cdots + z_N$ is equal to the expected value of $(N - n)$ times the expected value of a single z, i.e.,

$$(3:53) \quad E(z_{n+1} + \cdots + z_N) = E(N - n)E(z) = NE(z) - E(n)E(z).$$

From (3:51) and (3:53) it follows that

$$(3:54) \qquad E(z_1 + \cdots + z_n) - E(n)E(z) = 0$$

Hence

$$(3:55) \qquad E(n) = \frac{E(z_1 + \cdots + z_n)}{E(z)}$$

if $E(z) \neq 0$.

If θ is the true value of the parameter, then $E(n) = E_\theta(n)$ by the definition of the symbol $E_\theta(n)$. We shall denote by $E_\theta(z)$ the expected value $E(z)$ of z when θ is the true value of the parameter. If the excess of the probability ratio p_{1m}/p_{0m} over the boundaries A and B at the termination of the sequential process is neglected, the random variable $(z_1 + \cdots + z_n)$ can take only the values $\log A$ and $\log B$ with the probabilities $1 - L(\theta)$ and $L(\theta)$, respectively. Hence

$$(3:56) \qquad E(z_1 + \cdots + z_n) \sim L(\theta) \log B + [1 - L(\theta)] \log A$$

From (3:55) and (3:56) we obtain the approximation formula

$$(3:57) \qquad E_\theta(n) \sim \frac{L(\theta) \log B + [1 - L(\theta)] \log A}{E_\theta(z)}$$

In the preceding section we have computed explicitly the formula $L(\theta)$ for the binomial and normal case. Thus, to obtain the explicit formula for $E_\theta(n)$, we need only compute $E_\theta(z)$. In the binomial case, i.e., when $f(x, \theta) = \theta$ for $x = 1$ and $f(x, \theta) = 1 - \theta$ for $x = 0$, we have

$$(3:58) \quad E_\theta(z) = E_\theta\left[\log \frac{f(x, \theta_1)}{f(x, \theta_0)}\right] = \theta \log \frac{f(1, \theta_1)}{f(1, \theta_0)} + (1 - \theta) \log \frac{f(0, \theta_1)}{f(0, \theta_0)}$$

$$= \theta \log \frac{\theta_1}{\theta_0} + (1 - \theta) \log \frac{1 - \theta_1}{1 - \theta_0}$$

In the normal case, i.e., when

$$f(x, \theta) = \frac{1}{\sqrt{2\pi}\sigma} e^{-\frac{1}{2\sigma^2}(x-\theta)^2}$$

we have

(3:59) $z = \log \dfrac{f(x, \theta_1)}{f(x, \theta_0)} = \dfrac{1}{2\sigma^2} [2(\theta_1 - \theta_0)x + \theta_0{}^2 - \theta_1{}^2]$

Hence,

(3:60) $E_\theta(z) = \dfrac{1}{2\sigma^2} [2(\theta_1 - \theta_0)\theta + \theta_0{}^2 - \theta_1{}^2]$

3.6 Saving in the Number of Observations Effected by the Use of the Sequential Probability Ratio Test instead of the Current Test Procedure

In this section we shall assume that H_0 is the hypothesis that the random variable x under consideration is normally distributed with mean θ_0 and variance unity, while H_1 is the hypothesis that x is normally distributed with mean θ_1 and variance unity. We may assume without loss of generality that $\theta_0 < \theta_1$. We shall compare the expected number of observations required by the sequential probability ratio test of strength (α, β) for testing H_0 against H_1 with the fixed number of observations needed for the current most powerful test to attain the same strength (α, β).

We shall denote by $n(\alpha, \beta)$ the fixed number of observations required by the current test to attain the strength (α, β). The current most powerful test procedure for testing H_0 against H_1 is carried out as follows. The hypothesis H_0 is accepted if the arithmetic mean $\bar{x}$ of the observations $x_1, \cdots, x_n$ (the number n of observations is determined in advance) is less than or equal to a preassigned constant d, and H_0 is rejected (H_1 is accepted) if $\bar{x}$ exceeds d. The constant d and the fixed number n of observations are to be determined so that the test will have the required strength (α, β). For any given n and d the corresponding strength of the test can be determined as follows. Since $\bar{x} \leqq d$ is equivalent to the inequality $\sqrt{n}(\bar{x} - \theta_0) \leqq \sqrt{n}(d - \theta_0)$, the probability that $\bar{x} \leqq d$ is the same as the probability that $\sqrt{n}(\bar{x} - \theta_0) \leqq \sqrt{n}(d - \theta_0)$. The random variable $y = \sqrt{n}(\bar{x} - \theta_0)$ is normally distributed with mean 0 and variance unity if H_0 is true. Thus, the probability that $\bar{x} \leqq d$ when H_0 is true, i.e., the probability that we shall accept H_0 when H_0 is true, is equal to the probability that $y \leqq \sqrt{n}(d - \theta_0)$. We shall denote by $G(t)$ the probability that a normally distributed random variable with mean 0 and variance unity will take a value less than t, i.e.,

$$(3:61) \qquad G(t) = \frac{1}{\sqrt{2\pi}} \int_{-\infty}^{t} e^{-\frac{x^2}{2}} \, dx$$

Then the probability that we shall accept H_0 when H_0 is true is equal to $G[\sqrt{n}(d - \theta_0)]$. Since the probability that we shall accept H_0 when H_0 is true is $1 - \alpha$ by definition, we have

$$(3:62) \qquad G[\sqrt{n}(d - \theta_0)] = 1 - \alpha$$

To determine the value of β corresponding to given n and d, we shall write the inequality $\bar{x} \leq d$ in the equivalent form $\sqrt{n}(\bar{x} - \theta_1) \leq \sqrt{n}(d - \theta_1)$. By definition, β is the probability that we shall accept H_0 when H_1 is true. But the latter probability is the same as the probability that $\bar{x} \leq d$, i.e., that $\sqrt{n}(\bar{x} - \theta_1) \leq \sqrt{n}(d - \theta_1)$, when H_1 is true. But when H_1 is true this probability is equal to $G[\sqrt{n}(d - \theta_1)]$. Thus, we have

$$(3:63) \qquad G[\sqrt{n}(d - \theta_1)] = \beta$$

Hence, to obtain a test of the required strength (α, β), we have to choose the quantities n and d so that equations (3:62) and (3:63) are fulfilled. Let λ_0 be the value for which $G(\lambda_0) = 1 - \alpha$ and let λ_1 be the value for which $G(\lambda_1) = \beta$. The values λ_0 and λ_1 can be obtained from a table of the normal distribution. Then equations (3:62) and (3:63) can be written as

$$(3:64) \qquad \sqrt{n}(d - \theta_0) = \lambda_0$$

and

$$(3:65) \qquad \sqrt{n}(d - \theta_1) = \lambda_1$$

Subtracting equation (3:64) from equation (3:65) we obtain

$$(3:66) \qquad \sqrt{n}(\theta_0 - \theta_1) = \lambda_1 - \lambda_0$$

Thus,

$$(3:67) \qquad n = n(\alpha, \beta) = \frac{(\lambda_1 - \lambda_0)^2}{(\theta_0 - \theta_1)^2}$$

If this expression is not an integer, $n(\alpha, \beta)$ is the smallest integer in excess.

We shall now determine the expected number of observations required by the sequential probability ratio test of strength (α, β) and we shall compare it with the fixed number $n(\alpha, \beta)$ of observations required by the current test as given in formula (3:67). In the sequential test we shall use the approximation formulas for A and B, i.e., we shall let A and B equal $(1 - \beta)/\alpha$ and $\beta/(1 - \alpha)$, respectively, instead of the exact values $A(\alpha, \beta)$ and $B(\alpha, \beta)$, respectively. It has been shown in Section 3.2 that $(1 - \beta)/\alpha \geqq A(\alpha, \beta)$ and $\beta/(1 - \alpha)$ $\leqq B(\alpha, \beta)$. Thus, by letting $A = (1 - \beta)/\alpha$ and $B = \beta/(1 - \alpha)$ instead of using the exact values $A(\alpha, \beta)$ and $B(\alpha, \beta)$, we can only increase the number of observations required by the sequential test. Consequently, the saving effected by the sequential test of strength (α, β) as compared with the current test cannot be smaller than the saving which results from the sequential test obtained by using the approximation formulas $A = (1 - \beta)/\alpha$ and $B = \beta/(1 - \alpha)$.

We shall assume that $|\theta_1 - \theta_0|$ is small so that the approximation formula (3:57) for the expected value of n can be used. Since $L(\theta_0)$ $= 1 - \alpha$ and $L(\theta_1) = \beta$, we obtain from (3:57)

$$(3:68) \qquad E_1(n) = \frac{\beta \log B + (1 - \beta) \log A}{E_1(z)}$$

and

$$(3:69) \qquad E_0(n) = \frac{(1 - \alpha) \log B + \alpha \log A}{E_0(z)}$$

where $E_i(n)$ denotes the expected value of n when H_i is true $(i = 0, 1)$. As can easily be verified,

$$(3:70) \qquad E_1(z) = \tfrac{1}{2}(\theta_0 - \theta_1)^2$$

and

$$(3:71) \qquad E_0(z) = -\tfrac{1}{2}(\theta_0 - \theta_1)^2$$

From (3:67), (3:68), (3:69), (3:70), and (3:71) we obtain

$$(3:72) \qquad \frac{E_1(n)}{n(\alpha, \beta)} = \frac{2}{(\lambda_1 - \lambda_0)^2} [\beta \log B + (1 - \beta) \log A]$$

and

$$(3:73) \qquad \frac{E_0(n)}{n(\alpha, \beta)} = \frac{2}{(\lambda_1 - \lambda_0)^2} [-(1 - \alpha) \log B - \alpha \log A]$$

It is interesting to note that the ratios $\dfrac{E_1(n)}{n(\alpha, \beta)}$ and $\dfrac{E_0(n)}{n(\alpha, \beta)}$ are independent of the parameter values θ_0 and θ_1. The average saving of the sequential test as compared with the current test is $100 \left[1 - \dfrac{E_1(n)}{n(\alpha, \beta)} \right]$ per cent if H_1 is true, and $100 \left[1 - \dfrac{E_0(n)}{n(\alpha, \beta)} \right]$ per cent if H_0 is true.

In Table 1, panel A shows the value of $100\left[1 - \dfrac{E_1(n)}{n(\alpha,\beta)}\right]$, and panel B shows the value of $100\left[1 - \dfrac{E_0(n)}{n(\alpha,\beta)}\right]$, for several values of α and β. Because of the symmetry of the normal distribution, panel B is obtained from panel A simply by interchanging α and β.

TABLE 1

AVERAGE PERCENTAGE SAVING IN SIZE OF SAMPLE WITH SEQUENTIAL ANALYSIS, AS COMPARED WITH CURRENT MOST POWERFUL TEST FOR TESTING MEAN OF A NORMALLY DISTRIBUTED VARIATE

A. When alternative hypothesis is true:

β \ α	.01	.02	.03	.04	.05
.01	58	60	61	62	63
.02	54	56	57	58	59
.03	51	53	54	55	55
.04	49	50	51	52	53
.05	47	49	50	50	51

B. When null hypothesis is true:

β \ α	.01	.02	.03	.04	.05
.01	58	54	51	49	47
.02	60	56	53	50	49
.03	61	57	54	51	50
.04	62	58	55	52	50
.05	63	59	55	53	51

As the table shows, for the range of α and β from .01 to .05 (the range most frequently employed), the sequential test results in an average saving of at least 47 per cent in the necessary number of observations as compared with the current test. The true saving is slightly higher than shown in the table, since $E_i(n)$ ($i = 0, 1$) calculated under the condition that $A = (1 - \beta)/\alpha$ and $B = \beta/(1 - \alpha)$ is greater than $E_i(n)$ calculated under the condition that $A = A(\alpha, \beta)$ and $B = B(\alpha, \beta)$.

3.7   Lower Limit of the Probability That the Sequential Test Will Terminate with a Number of Trials Less Than or Equal to a Given Number

In Section A.6 an approximate formula [11] for the probability distribution of the number of observations required by the sequential test is derived in the case in which $z = \log \dfrac{f(x, \theta_1)}{f(x, \theta_0)}$ is normally distributed. It is pointed out that the same distribution function of n can be regarded as an approximation to the exact distribution even when z is not normally distributed, provided that the absolute value of $E(z)$ and the standard deviation of z are sufficiently small as compared with $\log A$ and $\log B$. Although the distribution of n given in Section A.6 could be used to determine the probability that $n \leqq n_0$ for any fixed integer n_0, we shall prefer to derive a lower limit for this probability by a different method for the following reasons.   (1) The computation of the lower limit given in this section is very simple, whereas the use of the distribution function given in Section A.6 would require laborious computations, since that distribution function has not yet been tabulated.   (2) If n_0 is fairly large and if α and β are small, as they usually are in practice, the lower bound given in this section will be fairly near the exact value.

For any given positive integer let $P_i(n \leqq n_0)$ denote the probability that $n \leqq n_0$ when H_i is true, i.e., when $\theta = \theta_i$ $(i = 0, 1)$.[12]   We want to derive a lower bound for $P_i(n \leqq n_0)$.   It will be assumed that n_0 is sufficiently large so that the sum $z_1 + \cdots + z_{n_0}$ may be regarded as normally distributed even when the distribution of z is not normal.[13]

If $\displaystyle\sum_{\alpha=1}^{n_0} z_\alpha \geqq \log A$, then we certainly have $n \leqq n_0$.   Similarly, if

$\displaystyle\sum_{\alpha=1}^{n_0} z_\alpha \leqq \log B$, we must have $n \leqq n_0$.   Hence

$$(3{:}74) \qquad P_1\left(\sum_{\alpha=1}^{n_0} z_\alpha \geqq \log A\right) \leqq P_1(n \leqq n_0)$$

and

$$(3{:}75) \qquad P_0\left(\sum_{\alpha=1}^{n_0} z_\alpha \leqq \log B\right) \leqq P_0(n \leqq n_0)$$

[11] See formulas (A:166), (A:183) and (A:194).

[12] In general, for any relation R we use the symbol $P_i(R)$ to denote the probability that R holds when H_i is true.

[13] According to well-known theorems in the theory of probability, the sum of a large number of independent random variables is nearly normally distributed under very general conditions.

The inequality $\sum_{\alpha=1}^{n_0} z_\alpha \geqq \log A$ can be written as

$$(3{:}76) \qquad \frac{\sum_{\alpha=1}^{n_0} z_\alpha - n_0 E_1(z)}{\sqrt{n_0}\sigma_1(z)} \geqq \frac{\log A - n_0 E_1(z)}{\sqrt{n_0}\sigma_1(z)}$$

where $\sigma_1(z)$ denotes the standard deviation of z when H_1 is true. The left-hand member of (3:76) is normally distributed with mean 0 and variance unity when H_1 is true. For any value λ we shall denote by $G(\lambda)$ the probability that a normally distributed random variable with mean 0 and variance unity will take a value less than λ. Thus, the probability that such a random variable takes a value $\geqq \lambda$ is given by $1 - G(\lambda)$. Hence the probability that (3:76) holds when H_1 is true is equal to $1 - G[\lambda_1(n_0)]$ where

$$(3{:}77) \qquad \lambda_1(n_0) = \frac{\log A - n_0 E_1(z)}{\sqrt{n_0}\sigma_1(z)}$$

But the probability that (3:76) holds when H_1 is true is equal to $P_1(\Sigma z_\alpha \geqq \log A)$. Thus,

$$(3{:}78) \qquad P_1\left(\sum_{\alpha=1}^{n_0} z_\alpha \geqq \log A\right) = 1 - G[\lambda_1(n_0)]$$

Because of (3:74), we obtain

$$1 - G[\lambda_1(n_0)] \leqq P_1(n \leqq n_0)$$

Thus, $1 - G[\lambda_1(n_0)]$ is a lower limit of the probability that $n \leqq n_0$, when H_1 is true.

To obtain a lower limit for $P_0(n \leqq n_0)$, we rewrite the inequality $\sum_{\alpha=1}^{n_0} z_\alpha \leqq \log B$ in the form

$$(3{:}79) \qquad \frac{\sum_{\alpha=1}^{n_0} z_\alpha - n_0 E_0(z)}{\sqrt{n_0}\sigma_0(z)} \leqq \frac{\log B - n_0 E_0(z)}{\sqrt{n_0}\sigma_0(z)} = \lambda_0(n_0), \quad \text{say}$$

where $\sigma_0(z)$ denotes the standard deviation of z when H_0 is true. Since the left-hand member of (3:79) is normally distributed with mean 0 and variance unity when H_0 is true, the probability that (3:79) holds when H_0 is true is equal to $G[\lambda_0(n_0)]$. Hence,

$$(3{:}80) \qquad P_0\left(\sum_{\alpha=1}^{n_0} z_\alpha \leqq \log B\right) = G[\lambda_0(n_0)]$$

Because of (3:75), we then have

$$(3:81) \qquad\qquad G[\lambda_0(n_0)] \leqq P_0(n \leqq n_0)$$

Thus $G[\lambda_0(n_0)]$ is a lower bound of the probability that $n \leqq n_0$ when H_0 is true.

When $\log A = \log(1 - \beta)/\alpha$ and $\log B = \log \beta/(1 - \alpha)$, Table 2 shows the values of the lower bounds of $P_0(n \leqq n_0)$ and $P_1(n \leqq n_0)$ corresponding to different pairs (α, β) and different values of n_0. In these calculations it has been assumed that the distribution under H_0 is a normal distribution with mean 0 and variance unity, and the distribution under H_1 is a normal distribution with mean θ and variance unity. For each pair (α, β) the value of θ has been determined from (3:67) so that the number of observations needed for the current most powerful test of strength (α, β) is equal to 1000.

TABLE 2

LOWER BOUND OF THE PROBABILITY THAT A SEQUENTIAL ANALYSIS WILL TERMINATE WITHIN VARIOUS NUMBERS OF TRIALS, WHEN THE MOST POWERFUL CURRENT TEST REQUIRES EXACTLY 1000 TRIALS

Number of Trials	$\alpha = .01$ and $\beta = .01$		$\alpha = .01$ and $\beta = .05$		$\alpha = .05$ and $\beta = .05$	
	Alternative hypothesis true	Null hypothesis true	Alternative hypothesis true	Null hypothesis true	Alternative hypothesis true	Null hypothesis true
1000	.910	.910	.799	.891	.773	.773
1200	.950	.950	.871	.932	.837	.837
1400	.972	.972	.916	.957	.883	.883
1600	.985	.985	.946	.972	.915	.915
1800	.991	.991	.965	.982	.938	.938
2000	.995	.995	.977	.989	.955	.955
2200	.997	.997	.985	.993	.967	.967
2400	.999	.999	.990	.995	.976	.976
2600	.999	.999	.994	.997	.982	.982
2800	1.00	1.00	.996	.998	.987	.987
3000	1.00	1.00	.997	.999	.990	.990

The probabilities given are lower bounds for the true probabilities. They relate to a test of the mean of a normally distributed variate, the difference between the null and alternative hypothesis being adjusted for each pair of values of α and β so that the number of trials required under the most powerful current test is exactly 1000.

3.8 Truncation of the Sequential Test Procedure

Although it is shown in Section A.1 that the probability is 1 that the sequential test procedure will eventually terminate, it is occasionally desirable to set a definite upper limit, say n_0, for the number of observations. This can be achieved by truncating the sequential process at $n = n_0$, i.e., by giving a new rule for the acceptance or rejection of H_0 at the n_0th trial if the sequential process did not lead to a final decision for $n \leq n_0$. A simple and reasonable rule for truncation at the n_0th trial seems to be the following: If the sequential probability ratio test does not lead to a final decision for $n \leq n_0$, accept H_0 at the n_0th trial when $\log B < \sum_{\alpha=1}^{n_0} z_\alpha \leq 0$, and reject H_0 when $0 < \sum_{\alpha=1}^{n_0} z_\alpha < \log A$.

By truncating the sequential process at the n_0th trial we shall, however, change the probabilities of errors of the first and second kinds. Let α and β be the probabilities of errors of the first and second kinds if the sequential test is not truncated. The effect of the truncation on α and β will, of course, depend on the value of n_0. The larger n_0, the smaller will be the effect of truncation on α and β. We shall denote the resulting probabilities of errors of the first and second kinds by $\alpha(n_0)$ and $\beta(n_0)$, respectively, if the sequential process is truncated at $n = n_0$. In this section we shall derive upper bounds for $\alpha(n_0)$ and $\beta(n_0)$.

To obtain an upper bound for $\alpha(n_0)$ we have to consider the cases in which the truncated process leads to the rejection of H_0, while the non-truncated process leads to the acceptance of H_0. Denote by $\rho_0(n_0)$ the probability under H_0 of obtaining a sample such that the truncated process leads to the rejection of H_0, while the non-truncated process leads to the acceptance of H_0. Then, we clearly have

(3:82)
$$\alpha(n_0) \leq \alpha + \rho_0(n_0)$$

The reason that in (3:82) the inequality sign holds instead of the equality sign is that there may be samples for which the truncated process leads to the acceptance of H_0, while the non-truncated process leads to the rejection of H_0. To obtain an upper bound for $\alpha(n_0)$, we merely need to derive an upper bound for $\rho_0(n_0)$. By definition, $\rho_0(n_0)$ is the probability under H_0 that for the successive observations z_1, z_2, $\cdots$, etc.. the following three conditions are simultaneously fulfilled:

(i) $\log B < \sum_{\alpha=1}^{n} z_\alpha < \log A$ for $n = 1, \cdots, n_0 - 1$

(*ii*)
$$0 < \sum_{\alpha=1}^{n_0} z_\alpha < \log A$$

(*iii*)   When the sequential process is continued beyond n_0, it terminates with the acceptance of H_0.

Denote by $\bar{p}_0(n_0)$ the probability under H_0 that condition (*ii*) will be fulfilled, i.e.,

(3:83)
$$\bar{p}_0(n_0) = P_0(0 < \sum_{\alpha=1}^{n_0} z_\alpha < \log A)$$

Since the probability that condition (*ii*) is fulfilled cannot be smaller than the probability that all three conditions are fulfilled simultaneously, we have

$$\bar{p}_0(n_0) \geqq p_0(n_0)$$

and, therefore,

(3:84)
$$\alpha(n_0) \leqq \alpha + \bar{p}_0(n_0)$$

Thus, $\alpha + \bar{p}_0(n_0)$ is an upper bound for $\alpha(n_0)$, which can easily be computed, as will be shown later. To obtain an upper bound for $\beta(n_0)$ we shall denote by $\rho_1(n_0)$ the probability (under H_1) that the successive observations will be such that the truncated process leads to the acceptance of H_0, while the non-truncated process leads to the rejection of H_0. In other words, $\rho_1(n_0)$ is the probability under H_1 that the successive observations will satisfy the following three conditions simultaneously:

(*i*)
$$\log B < \sum_{\alpha=1}^{n} z_\alpha < \log A \quad \text{for } n = 1, \cdots, n_0 - 1$$

(*ii*)
$$\log B < \sum_{\alpha=1}^{n_0} z_\alpha \leqq 0$$

(*iii*)   If the process is continued beyond the n_0th trial, it terminates with the acceptance of H_1.

Clearly

(3:85)
$$\beta(n_0) \leqq \beta + \rho_1(n_0)$$

Since it is difficult to determine the value of $\rho_1(n_0)$, we shall derive a simple upper bound for it. Let $\bar{p}_1(n_0)$ be the probability under H_1 that condition (*ii*) is fulfilled, i.e.,

(3:86)
$$\bar{p}_1(n_0) = P_1(\log B < \sum_{\alpha=1}^{n_0} z_\alpha \leqq 0)$$

Then $\bar{\rho}_1(n_0) \geqq \rho_1(n_0)$ and we have

$$(3:87) \qquad\qquad \beta(n_0) \leqq \beta + \bar{\rho}_1(n_0)$$

We shall now show how $\bar{\rho}_0(n_0)$ and $\bar{\rho}_1(n_0)$ can be computed. We shall assume that n_0 is sufficiently large so that $z_1 + \cdots + z_{n_0}$ may be regarded as a normally distributed variable. When H_i is true ($i = 0, 1$) the expected value of $z_1 + \cdots + z_{n_0}$ is equal to $n_0 E_i(z)$ and the standard deviation of $z_1 + \cdots + z_{n_0}$ is equal to $\sqrt{n_0}\sigma_i(z)$ where $\sigma_i(z)$ denotes the standard deviation of z when H_i is true. To compute $\bar{\rho}_0(n_0)$, we shall write the inequality $0 < \sum_{\alpha=1}^{n_0} z_\alpha < \log A$ in the following form:

$$(3:88) \qquad \frac{-n_0 E_0(z)}{\sqrt{n_0}\sigma_0(z)} < \frac{z_1 + \cdots + z_{n_0} - n_0 E_0(z)}{\sqrt{n_0}\sigma_0(z)} < \frac{\log A - n_0 E_0(z)}{\sqrt{n_0}\sigma_0(z)}$$

Let

$$(3:89) \qquad \nu_1 = \frac{-n_0 E_0(z)}{\sqrt{n_0}\sigma_0(z)} \quad \text{and} \quad \nu_2 = \frac{\log A - n_0 E_0(z)}{\sqrt{n_0}\sigma_0(z)}$$

Since the middle term in (3:88) is normally distributed with zero mean and unit variance when H_0 is true, the probability that (3:88) is fulfilled when H_0 is true is equal to $G(\nu_2) - G(\nu_1)$ where $G(\nu)$ denotes the probability that a normally distributed variable with mean 0 and variance unity will take a value $< \nu$. Thus,

$$(3:90) \qquad\qquad \bar{\rho}_0(n_0) = G(\nu_2) - G(\nu_1)$$

To compute $\bar{\rho}_1(n_0)$, we shall write the inequality $\log B < \sum_{\alpha=1}^{n_0} z_\alpha \leqq 0$ in the following form:

$$(3:91) \qquad \frac{\log B - n_0 E_1(z)}{\sqrt{n_0}\sigma_1(z)} < \frac{z_1 + \cdots + z_{n_0} - n_0 E_1(z)}{\sqrt{n_0}\sigma_1(z)} \leqq \frac{-n_0 E_1(z)}{\sqrt{n_0}\sigma_1(z)}$$

Let

$$(3:92) \qquad \nu_3 = \frac{\log B - n_0 E_1(z)}{\sqrt{n_0}\sigma_1(z)} \quad \text{and} \quad \nu_4 = \frac{-n_0 E_1(z)}{\sqrt{n_0}\sigma_1(z)}$$

Since the middle term in (3:91) is normally distributed with mean 0 and variance unity when H_1 is true, the probability (under H_1) that (3:91) holds is equal to $G(\nu_4) - G(\nu_3)$. Hence,

$$(3:93) \qquad\qquad \bar{\rho}_1(n_0) = G(\nu_4) - G(\nu_3)$$

Our results can thus be summarized as follows:

$$(3:94) \qquad \alpha(n_0) \leqq \alpha + G(\nu_2) - G(\nu_1)$$

and

$$(3:95) \qquad \beta(n_0) \leqq \beta + G(\nu_4) - G(\nu_3)$$

where ν_1, ν_2, ν_3, and ν_4 are given in (3:89) and (3:92). These upper bounds may considerably exceed $\alpha(n_0)$ and $\beta(n_0)$, respectively. It would be desirable to find closer limits.

Table 3 shows the values of the upper bounds of $\alpha(n_0)$ and $\beta(n_0)$ given in (3:94) and (3:95) corresponding to different pairs (α, β) and different values of n_0. In these calculations we have put $\log A =$

TABLE 3

EFFECT ON RISKS OF ERROR OF TRUNCATING A SEQUENTIAL ANALYSIS
AT A PREDETERMINED NUMBER OF TRIALS

Number of Trials	$\alpha = .01$ and $\beta = .01$		$\alpha = .01$ and $\beta = .05$		$\alpha = .05$ and $\beta = .05$	
	Upper bound of effective α	Upper bound of effective β	Upper bound of effective α	Upper bound of effective β	Upper bound of effective α	Upper bound of effective β
1000	.020	.020	.033	.070	.095	.095
1200	.015	.015	.024	.063	.082	.082
1400	.013	.013	.019	.058	.072	.072
1600	.012	.012	.016	.055	.066	.066
1800	.011	.011	.014	.053	.062	.062
2000	.010	.010	.012	.052	.058	.058
2200	.010	.010	.012	.051	.056	.056
2400	.010	.010	.011	.051	.055	.055
2600	.010	.010	.011	.051	.053	.053
2800	.010	.010	.010	.050	.053	.053
3000	.010	.010	.010	.050	.052	.052

If the sequential analysis is based on the values α and β shown, but a decision is made at n_0 trials even when the normal sequential criteria would require a continuation of the process, the realized values $\alpha(n_0)$ and $\beta(n_0)$ will not exceed the tabular entries. The table relates to a test of the mean of a normally distributed variate, the difference between the null and alternative hypotheses being adjusted for each pair (α, β) so that the number of trials required by the current test is 1000.

$\log (1 - \beta)/\alpha$ and $\log B = \log \beta/(1 - \alpha)$, and assumed that the distribution under H_0 is normal with mean 0 and variance unity, and the distribution under H_1 is normal with mean θ and variance unity. For each pair (α, β) the value of θ has been determined so that the number of observations required by the current most powerful test of strength (α, β) is equal to 1000.

It seems to the author that the upper limits given in (3:94) and (3:95) are considerably above the true $\alpha(n_0)$ and $\beta(n_0)$, respectively, when n_0 is not much higher than the value of n needed for the current most powerful test.

3.9 Increase in the Expected Number of Observations Caused by Replacing the Exact Values $A(\alpha, \beta)$ and $B(\alpha, \beta)$ by $(1 - \beta)/\alpha$ and $\beta/(1 - \alpha)$, Respectively

The quantities $A(\alpha, \beta)$ and $B(\alpha, \beta)$ denote the values of A and B for which the probabilities of errors of the first and second kinds associated with the sequential probability ratio test are exactly α and β, respectively. In Section 3.3 it has been recommended that $A(\alpha, \beta)$ and $B(\alpha, \beta)$ be replaced by $a(\alpha, \beta) = (1 - \beta)/\alpha$ and $b(\alpha, \beta) = \beta/(1 - \alpha)$, respectively. This may slightly increase the expected number of observations, since $a(\alpha, \beta) \geq A(\alpha, \beta)$ and $b(\alpha, \beta) \leq B(\alpha, \beta)$.[14] The present section gives estimates of the amount of such increase in the expected number of observations.

In Section 3.5 the following approximation formula has been obtained for the expected number of observations:

$$(3:96) \qquad E_\theta(n) \sim \frac{L(\theta) \log B + [1 - L(\theta)] \log A}{E_\theta(z)}$$

Since $L(\theta_0) = 1 - \alpha$ and $L(\theta_1) = \beta$, we obtain from (3:96)

$$(3:97) \qquad E_0(n) \sim \frac{(1 - \alpha) \log B + \alpha \log A}{E_0(z)}$$

and

$$(3:98) \qquad E_1(n) \sim \frac{\beta \log B + (1 - \beta) \log A}{E_1(z)}$$

$E_i(n)$ denotes the expected value of n when θ_i is true.

---

[14] See inequalities (3:21) and (3:22).

Thus, the changes $\Delta E_0(n)$ and $\Delta E_1(n)$ in the expected values $E_0(n)$ and $E_1(n)$ caused by using $a(\alpha, \beta)$ and $b(\alpha, \beta)$ instead of $A(\alpha, \beta)$ and $B(\alpha, \beta)$, respectively, are given by

$$(3{:}99) \qquad \Delta E_0(n) \sim \frac{\left\{ \begin{matrix} (1 - \alpha)[\log b(\alpha, \beta) - \log B(\alpha, \beta)] + \\ \alpha[\log a(\alpha, \beta) - \log A(\alpha, \beta)] \end{matrix} \right\}}{E_0(z)}$$

$$= \frac{(1 - \alpha) \log \dfrac{b(\alpha, \beta)}{B(\alpha, \beta)} + \alpha \log \dfrac{a(\alpha, \beta)}{A(\alpha, \beta)}}{E_0(z)}$$

and

$$(3{:}100) \qquad \Delta E_1(n) \sim \frac{\left\{ \begin{matrix} \beta[\log b(\alpha, \beta) - \log B(\alpha, \beta)] + \\ (1 - \beta)[\log a(\alpha, \beta) - \log A(\alpha, \beta)] \end{matrix} \right\}}{E_1(z)}$$

$$= \frac{\beta \log \dfrac{b(\alpha, \beta)}{B(\alpha, \beta)} + (1 - \beta) \log \dfrac{a(\alpha, \beta)}{A(\alpha, \beta)}}{E_1(z)}$$

Formulas (3:99) and (3:100) are, of course, approximation formulas, since (3:97) and (3:98) are approximations. However, if the error in the formulas (3:97) and (3:98), i.e., if the differences

$$(3{:}101a) \qquad E_0(n) - \frac{(1 - \alpha) \log B + \alpha \log A}{E_0(z)}$$

and

$$(3{:}101b) \qquad E_1(n) - \frac{\beta \log B + (1 - \beta) \log A}{E_1(z)}$$

were exactly independent of the quantities A and B, then in (3:99) and (3:100) the equality sign would hold exactly. It can be shown that small changes in A and B affect the differences (3:101) exceedingly little, and, therefore, (3:99) and (3:100) are very close approximations.

We shall derive upper bounds for the right-hand members of (3:99) and (3:100). Since $E_0(z)$ and $\log \dfrac{b(\alpha, \beta)}{B(\alpha, \beta)}$ are negative,[15] while $\log \dfrac{a(\alpha, \beta)}{A(\alpha, \beta)}$ is positive, we have

---

[15] It is remarked at the end of Section A.2.1 that $E(z)$ and a certain quantity h_0 defined there have opposite signs. Since $h_0 = 1$ if H_0 is true, and $h_0 = -1$ if H_1 is true, it follows that $E_0(z) < 0$ and $E_1(z) > 0$.

$$\text{(3:102)} \quad \frac{(1-\alpha)\log\dfrac{b(\alpha,\beta)}{B(\alpha,\beta)} + \alpha\log\dfrac{a(\alpha,\beta)}{A(\alpha,\beta)}}{E_0(z)} < \frac{(1-\alpha)\log\dfrac{b(\alpha,\beta)}{B(\alpha,\beta)}}{E_0(z)}$$

$$< \frac{1}{E_0(z)}\log\frac{b(\alpha,\beta)}{B(\alpha,\beta)}$$

Similarly, since $E_1(z)$ and $\log\dfrac{a(\alpha,\beta)}{A(\alpha,\beta)}$ are positive, while $\log\dfrac{b(\alpha,\beta)}{B(\alpha,\beta)}$ is negative, we have

$$\text{(3:103)} \quad \frac{\beta\log\dfrac{b(\alpha,\beta)}{B(\alpha,\beta)} + (1-\beta)\log\dfrac{a(\alpha,\beta)}{A(\alpha,\beta)}}{E_1(z)} < \frac{(1-\beta)\log\dfrac{a(\alpha,\beta)}{A(\alpha,\beta)}}{E_1(z)}$$

$$< \frac{1}{E_1(z)}\log\frac{a(\alpha,\beta)}{A(\alpha,\beta)}$$

Thus, for all practical purposes $\dfrac{1}{E_1(z)}\log\dfrac{a(\alpha,\beta)}{A(\alpha,\beta)}$ is an upper bound for $\Delta E_1(n)$ and $\dfrac{1}{E_0(z)}\log\dfrac{b(\alpha,\beta)}{B(\alpha,\beta)}$ is an upper bound for $\Delta E_0(n)$. The exact values $A(\alpha,\beta)$ and $B(\alpha,\beta)$ not being known, we cannot yet use these limits. Since $E_1(z) > 0$, an upper limit of $\dfrac{1}{E_1(z)}\log\dfrac{a(\alpha,\beta)}{A(\alpha,\beta)}$ is obtained by substituting for $\dfrac{a(\alpha,\beta)}{A(\alpha,\beta)}$ an upper bound of $\dfrac{a(\alpha,\beta)}{A(\alpha,\beta)}$. Similarly, since $E_0(z) < 0$, an upper limit of $\dfrac{1}{E_0(z)}\log\dfrac{b(\alpha,\beta)}{B(\alpha,\beta)}$ can be obtained by substituting for $\dfrac{b(\alpha,\beta)}{B(\alpha,\beta)}$ a lower bound of $\dfrac{b(\alpha,\beta)}{B(\alpha,\beta)}$.

From equations (A:29) and (A:30) in the Appendix one can derive the following inequalities:

$$\text{(3:104)} \quad \frac{a(\alpha,\beta)}{A(\alpha,\beta)} \leqq \delta_{\theta_0}$$

and

$$\text{(3:105)} \quad \frac{b(\alpha,\beta)}{B(\alpha,\beta)} \geqq \eta_{\theta_0}$$

where the quantities δ_θ and η_θ are defined by equations (A:27) and

(A:28).[16] The quantities δ_θ and η_θ have been explicitly computed for binomial and normal distributions.

Thus, we arrive at the following result: For all practical purposes we may regard $(\log \delta_{\theta_0})/E_1(z)$ as an upper bound for $\Delta E_1(n)$ and $(\log \eta_{\theta_0})/E_0(z)$ as an upper bound for $\Delta E_0(n)$.

TABLE 4

INCREASE IN EXPECTED NUMBER OF OBSERVATIONS RESULTING FROM APPROXIMATIONS IN CRITERIA FOR TERMINATING A SEQUENTIAL PROCESS

Number of Observations Needed for the Current Most Powerful Test	$\alpha = .01$ $\beta = .01$	$\alpha = .01$ $\beta = .05$	$\alpha = .05$ $\beta = .05$
10	1.1	1.3	1.6
30	1.9	2.2	2.7
100	3.4	4.0	4.9
200	4.9	5.7	6.9
500	7.7	9.0	10.9
1000	10.8	12.7	15.4

The tabular entries may, for practical purposes, be treated as upper bounds of the exact increases. The table relates to a test of the mean of a normally distributed variate, the difference between the null and alternative hypotheses being adjusted for each pair of values of α and β so that the number of trials required under the best current test is as shown in the left-hand column.

---

[16] This can be seen as follows: Substituting $A(\alpha, \beta)$ for A, $B(\alpha, \beta)$ for B, and θ_0 for θ, we obtain from (A:29) and (A:30)

$$[B(\alpha, \beta)]^{h(\theta_0)}\eta_{\theta_0} \leqq E_{\theta_0}{}^*$$

and

$$E_{\theta_0}{}^{**} \leqq [A(\alpha, \beta)]^{h(\theta_0)}\delta_{\theta_0}$$

Since we let $A = A(\alpha, \beta)$ and $B = B(\alpha, \beta)$, we have $L(\theta_0) = 1 - \alpha$ and $L(\theta_1) = \beta$. It follows from this and the two equations which are obtained from (A:18) by substituting θ_0 and θ_1 for θ that

$$E_{\theta_0}{}^* = \frac{\beta}{1 - \alpha} = b(\alpha, \beta) \quad \text{and} \quad E_{\theta_0}{}^{**} = \frac{1 - \beta}{\alpha} = a(\alpha, \beta)$$

Since $h(\theta_0) = 1$, we obtain

$$B(\alpha, \beta)\eta_{\theta_0} \leqq b(\alpha, \beta) \quad \text{and} \quad a(\alpha, \beta) \leqq A(\alpha, \beta)\delta_{\theta_0}$$

from which (3:104) and (3:105) follow.

As an example, consider the case in which the distribution under H_0 is normal with zero mean and unit variance, and the distribution under H_1 is normal with mean θ and variance unity. Since for the normal distribution $\eta_\theta = 1/\delta_\theta$ [see equation (A:51)] and $-E_0(z) = E_1(z)$, the upper bound of $\Delta E_0(n)$ is the same as the upper bound of $\Delta E_1(n)$. This upper bound depends only on the value of θ. For any pair (α, β) and for any positive integer m there exists exactly one value of θ such that m observations are needed for the current most powerful test of strength (α, β). Thus, with each integer m and pair (α, β) there is associated exactly one value of θ. Table 4 shows the common upper bound of $\Delta E_0(n)$ and $\Delta E_1(n)$ calculated for values of θ corresponding to different pairs (α, β) and integers m.

Chapter 4. OUTLINE OF A THEORY OF SEQUENTIAL TESTS OF SIMPLE AND COMPOSITE HYPOTHESES AGAINST A SET OF ALTERNATIVES

In Chapter 3 we were concerned mainly with the theoretical case of testing a simple hypothesis H_0 against a single alternative H_1. In problems arising in applications, the unknown parameter, or parameters, can usually take infinitely many values. In this chapter we shall discuss sequential tests of simple and composite hypotheses against infinitely many alternatives.

4.1 Tests of Simple Hypotheses

4.1.1 Introductory Remarks

A simple hypothesis has been defined as a statement which specifies completely the values of all the unknown parameters. We should like to make some remarks concerning the conditions under which a test of a simple hypothesis is meaningful and appropriate. For this purpose it will be sufficient to consider the case in which there is only one unknown parameter θ involved in the distribution of the random variable x under consideration. A simple hypothesis is then a statement that θ is equal to some specified value θ_0.

In applications the problem of testing a hypothesis usually arises as follows: There are two alternative courses of action, say action 1 and action 2, between which a decision is to be made, and the preference for one or the other action depends on the value of the parameter θ. Let ω denote the set of all values of θ for which action 1 is preferred to action 2; then action 2 is preferred to action 1 for all values θ outside ω.[1] Let H_ω be the hypothesis that θ is contained in ω. Then the problem of deciding between the two courses of action can be formulated as the problem of testing the hypothesis H_ω. If H_ω is accepted we take action 1 and if H_ω is rejected we take action 2. If the degree of preference for one or the other action varies continuously with the value of θ, the set ω cannot consist of a single value θ_0. In fact, if ω were to contain only the single value θ_0, it would mean that we prefer action 1 when $\theta = \theta_0$ and we prefer action 2 for any $\theta \neq \theta_0$, no matter

---

[1] For values θ on the boundary of ω it will usually be inconsequential which action is taken.

how near θ is to θ_0. Thus, we would have a discontinuity in our preference scale at $\theta = \theta_0$.

We see that the problem of testing a simple hypothesis arises, strictly
speaking, only if there is a discontinuity in our preference scale for
actions 1 and 2. While a discontinuity in the preference scale is, of
course, possible, it will occur rather seldom. A discontinuity in the
preference scale may occur, for example, if we want to test the validity
of some hypothetical scientific theory which implies that the parameter θ must have a specified value θ_0. In such a case any deviation of
the value of θ from θ_0, no matter how small, is of importance, since it
invalidates the hypothetical theory in question.

Whenever the degree of preference for one or the other action varies
continuously with the value of θ, the hypothesis to be tested will have
to be, strictly speaking, a composite one. Nevertheless, frequently it
will be expedient to approximate the composite hypothesis by a simple
one, since the latter is usually a simpler problem to treat. As an illustration, consider the following example: Suppose that the hardness x
of a material varies from unit to unit and is normally distributed with
a known variance. The mean value θ of x is, however, unknown. Suppose that θ_0 is considered to be the most desirable value of θ and the
material is considered less desirable the greater $\left| \theta - \theta_0 \right|$. Let action
1 be acceptance of the material and action 2, rejection of the material.
Preference for acceptance is strongest when $\theta = \theta_0$. The preference
for acceptance will decrease steadily as $\left| \theta - \theta_0 \right|$ increases. There will
be a positive value δ such that for $\left| \theta - \theta_0 \right| > \delta$ rejection of the mate-
rial is preferred and the degree of preference for rejection increases
with increasing value of $\left| \theta - \theta_0 \right|$ in the domain $\left| \theta - \theta_0 \right| > \delta$. If
$\left| \theta - \theta_0 \right| = \delta$, i.e., if the quality of the product is just on the margin,
neither action is preferable to the other. In such a situation the proper
hypothesis to be tested is the composite hypothesis that $\left| \theta - \theta_0 \right| \leqq \delta$.
However, if δ is small, the composite hypothesis may be replaced for
practical purposes by the simple hypothesis that $\theta = \theta_0$. The test of
the hypothesis that $\theta = \theta_0$ will have nearly the same operating charac-
teristic function as the test of the hypothesis that $\left| \theta - \theta_0 \right| \leqq \delta$, for
the following reasons. To test the hypothesis that $\left| \theta - \theta_0 \right| \leqq \delta$ we
subdivide the θ-axis into three zones: zone of preference for acceptance,
zone of preference for rejection, and zone of indifference. As explained
in Section 2.3.1, the zone of preference for acceptance consists of all
values θ for which acceptance is strongly preferred, i.e., for which the
rejection of the material is considered an error of practical importance.
Similarly, the zone of preference for rejection consists of all those values
θ for which rejection is strongly preferred, whereas for values θ in the

indifferent zone the preference for one action over the other is only slight and we do not care particularly which action is taken. In our example the three zones may reasonably be defined as follows. We select two positive values $\delta_0 < \delta$ and $\delta_1 > \delta$. The zone of preference for acceptance is given by $|\theta - \theta_0| \leqq \delta_0$, the zone of preference for rejection by $|\theta - \theta_0| \geqq \delta_1$, and the zone of indifference by $\delta_0 < |\theta - \theta_0| < \delta_1$. The test procedure will then be constructed so that the probability of rejection will not exceed a preassigned value α whenever θ is in the zone of preference for acceptance, and the probability of acceptance will not exceed a preassigned value β whenever θ is in the zone of preference for rejection.[2]   Now if we replace the original composite hypothesis by the simple hypothesis that $\theta = \theta_0$, the zone of preference for acceptance will consist of the single value $\theta = \theta_0$. The zone of preference for rejection may be defined, as before, by $|\theta - \theta_0| \geqq \delta_1$. The zone of indifference is then given by $0 < |\theta - \theta_0| < \delta_1$. The test procedure for testing that $\theta = \theta_0$ will then satisfy the requirement that the probability of rejecting the hypothesis is α when $\theta = \theta_0$ and the probability of accepting the hypothesis does not exceed β whenever $|\theta - \theta_0| \geqq \delta_1$. If δ_0 is very small, the test of the hypothesis that $\theta = \theta_0$ will satisfy the requirements imposed on the test of the original composite hypothesis with close approximation, since the probability of rejecting the hypothesis will be nearly equal to α for values θ in a sufficiently small neighborhood of θ_0. Thus, for practical purposes we may replace the original composite hypothesis by the simple hypothesis that $\theta = \theta_0$.

As we have seen, a test of a simple hypothesis will occur in applications in two cases: (1) when there is a discontinuity in the preference scale and the problem calls for testing a simple hypothesis in the strict sense (these cases are rare); (2) when the problem is such that it calls for testing a composite hypothesis and it is approximated by a simple hypothesis merely for the sake of simplicity.

In terms of the zones of preference for acceptance, of preference for rejection, and of indifference, the simple hypothesis may be characterized by the condition that the zone of preference for acceptance consists of a single point.

4.1.2   Test of a Simple Hypothesis against One-Sided Alternatives

We shall discuss here the simple case in which there is only one unknown parameter θ and the hypothesis that $\theta = \theta_0$ is tested against alternative values of θ which lie on one side of θ_0, say $> \theta_0$. In other words, only values of $\theta > \theta_0$ are considered admissible alternatives to

---

[2] In this connection see Section 2.3.2.

the hypothesis to be tested. In this case the zone of preference for acceptance consists of the single value θ_0. The degree of preference for rejection of the hypothesis will generally increase with increasing value of θ in the domain $\theta > \theta_0$. It will, therefore, be possible to find a value $\theta_1 > \theta_0$ such that the acceptance of the hypothesis is considered an error of practical importance whenever $\theta \geq \theta_1$, while for values $\theta > \theta_0$ but $< \theta_1$ the acceptance of the hypothesis is an error of no particular practical consequence. Thus, the zone of preference for rejection may be defined by $\theta \geq \theta_1$, and the zone of indifference by $\theta_0 < \theta < \theta_1$.

According to Section 2.3.2 we shall impose the following requirements on the OC function of the test. The probability that the hypothesis will be rejected should be equal to a preassigned value α when $\theta = \theta_0$. The probability of accepting the hypothesis should not exceed a preassigned value β whenever $\theta \geq \theta_1$.

In most of the important cases occurring in practice, such as when x has a normal, binomial, or Poisson distribution, and so on, the sequential probability ratio test of strength (α, β) for testing the hypothesis that $\theta = \theta_0$ against the single alternative θ_1 will satisfy the imposed requirements, since the probability of an error of the second kind will decrease steadily with increasing values of θ in the domain $\theta \geq \theta_1$. Thus, in all these cases the sequential probability ratio test for testing the hypothesis that $\theta = \theta_0$ against a properly chosen alternative θ_1 provides a satisfactory solution to our problem.

The case in which the alternative values of θ are restricted to values $\theta < \theta_0$ instead of values $> \theta_0$ is entirely analogous and need not be discussed separately.

4.1.3 Test of a Simple Hypothesis with No Restrictions on the Alternative Values of the Unknown Parameters

In this section we shall deal with the following general problem: The distribution of x involves k unknown parameters $\theta_1, \cdots, \theta_k$ and the hypothesis H_0 to be tested is that $\theta_1, \cdots, \theta_k$ are equal to some specified values $\theta_1^0, \cdots, \theta_k^0$, respectively. The set of k parameters $(\theta_1, \cdots, \theta_k)$ will be denoted by θ without any subscript and will be referred to as a parameter point. The use of a superscript to the letter θ, such as θ^0 or θ^1, etc., will indicate that a particular parameter point is meant. Our hypothesis H_0 can thus be expressed by stating that the unknown parameter point θ is equal to the particular parameter point θ^0.

As we have seen in the preceding section, the zone of preference for acceptance consists of the single parameter point θ^0. Denote the zone of preference for rejection by ω_r. This will usually be the set of all points θ whose "distance" (defined in some sense) from θ^0 is greater than or equal to some given positive value. The requirements imposed on the OC function of the test, as formulated in Section 2.3.2, can then be stated as follows: The probability that H_0 will be rejected when $\theta = \theta^0$ should be equal to a preassigned value α and the probability that H_0 will be accepted should not exceed a preassigned value β for any parameter point θ in the zone ω_r.

Before we discuss the problem of constructing a proper sequential test satisfying the above requirements, we shall consider the problem of finding a proper test procedure satisfying the following modified requirements. For any θ in ω_r let $\beta(\theta)$ denote the probability that H_0 will be accepted when θ is the true parameter point. Thus $\beta(\theta)$ is the probability of an error of the second kind when θ is true. Our original requirement was that $\beta(\theta)$ should not exceed a preassigned value β for all θ in ω_r. Instead we shall now require that the weighted average of $\beta(\theta)$, weighted with a given weight function $w(\theta)$, should be equal to β, i.e.,

$$(4\!:\!1) \qquad \int_{\omega_r} \beta(\theta) w(\theta)\, d\theta = \beta$$

where $w(\theta) \geqq 0$ for all θ in ω_r and [3]

$$(4\!:\!2) \qquad \int_{\omega_r} w(\theta)\, d\theta = 1$$

The requirement that the probability of rejecting H_0 when H_0 is true be equal to a preassigned α is maintained as before. A proper sequential test procedure satisfying these modified requirements can easily be constructed. Let p_{0n} be the probability distribution of the sample $(x_1, \cdots, x_n)$ when H_0 is true, i.e.,

$$(4\!:\!3) \quad p_{0n} = f(x_1, \theta_1^0, \cdots, \theta_k^0) f(x_2, \theta_1^0, \cdots, \theta_k^0) \cdots f(x_n, \theta_1^0, \cdots, \theta_k^0)$$

Furthermore, let p_{1n} be defined by

$$(4\!:\!4) \qquad p_{1n} = \int_{\omega_r} f(x_1, \theta_1, \cdots, \theta_k) \cdots f(x_n, \theta_1, \cdots, \theta_k) w(\theta)\, d\theta$$

Thus, p_{1n} is a weighted average of the probability distribution functions $f(x_1, \theta_1, \cdots, \theta_k) \cdots f(x_n, \theta_1, \cdots, \theta_k)$ corresponding to various parameter points θ in ω_r. As such, p_{1n} itself is a probability distribu-

[3] The weight function $w(\theta)$ may also be discrete. A single formula valid for both, continuous and discrete, weight functions could be given by using Stieltje's integrals in (4:1) and (4:2).

tion function of the sample $(x_1, \cdots, x_n)$.[4] Let H_1 denote the hypothesis that the distribution of the sample $(x_1, \cdots, x_n)$ is given by p_{1n} defined in (4:4). Then H_1 is a simple hypothesis, since it specifies completely the distribution. Consider the sequential probability ratio test of strength (α, β) for testing H_0 against the simple alternative hypothesis H_1. This procedure is given as follows. Reject H_0 if

$$(4:5) \qquad\qquad \frac{p_{1n}}{p_{0n}} \geqq A$$

accept H_0 if

$$(4:6) \qquad\qquad \frac{p_{1n}}{p_{0n}} \leqq B$$

and take an additional observation if

$$(4:7) \qquad\qquad B < \frac{p_{1n}}{p_{0n}} < A$$

The expressions p_{0n} and p_{1n} are given by (4:3) and (4:4), respectively, and the constants A and B are to be chosen so that the test will have the required strength (α, β). As we have seen in Section 3.3, for most practical purposes we may use the approximation formulas $A = (1 - \beta)/\alpha$ and $B = \beta/(1 - \alpha)$.[5]

The sequential probability ratio test defined by (4:5), (4:6), and (4:7) can be shown to satisfy the relation (4:1). Thus, this probability ratio test may be regarded as a satisfactory solution to our problem if our requirement is that the probability of an error of the first kind should be α and that $\beta(\theta)$ should satisfy (4:1).

In practical problems, however, it seems more reasonable to maintain the original requirements. That is to say, we shall want a test procedure such that the probability $\beta(\theta)$ of accepting H_0 does not exceed β for all parameter points θ in the zone ω_r, and the probability is α that we shall reject H_0 when $\theta = \theta^0$. There are, in general, infinitely many sequential tests which satisfy these requirements, and we want to select one for which the expected number of observations is as small as possible.

----

[4] The distribution of the sample $(x_1, \cdots, x_n)$ will be precisely given by p_{1n} if we assume that θ in ω_r has a probability distribution given by the density function $w(\theta)$.

[5] Although the successive observations $x_1, x_2, \cdots$, etc., are not independent when H_1 is true (p_{1n} cannot be represented as a product of n factors where the αth factor depends only on x_α), the results and conclusion in Sections 3.2 and 3.3 remain valid, as pointed out in Section 3.2.

Although a thorough investigation of this problem has not yet been made, the following approach may perhaps be reasonable. First we restrict ourselves to the class C of sequential probability ratio tests based on the ratio p_{1n}/p_{0n}, where p_{0n} is given by (4:3) and p_{1n} by (4:4), corresponding to an arbitrary non-negative weight function $w(\theta)$ satisfying (4:2).[6] Thus, the class C contains at least as many tests as there are possible weight functions $w(\theta)$ satisfying (4:2). A test in class C is uniquely determined by choosing a particular weight function $w(\theta)$ and particular values for A and B. The test procedure is then carried out in the usual way. H_0 is accepted if $p_{1n}/p_{0n} \leqq B$, H_0 is rejected if $p_{1n}/p_{0n} \geqq A$, and an additional observation is made if $B < p_{1n}/p_{0n} < A$. The restriction to the class C of sequential tests is suggested by the fact that we have been led to these tests by the requirement that some weighted average of the probabilities of errors of the second kind be equal to a given value β.

Accepting the restriction that the sequential test should be a member of the class C, we still need a principle for choosing the weight function $w(\theta)$. Suppose that the quantities A and B have already been determined. Let us then examine what would be a reasonable choice of $w(\theta)$. After A and B have been chosen, the probability α of making an error of the first kind is also determined for practical purposes and the choice of $w(\theta)$ will not affect it.[7] Thus, the choice of $w(\theta)$ will affect only $\beta(\theta)$. A weight function $w(\theta)$ may be regarded the more favorable the smaller the maximum value of $\beta(\theta)$ with respect to θ (θ is, of course, restricted to points in ω_r). Thus, the following choice of $w(\theta)$ seems reasonable: *For given values of A and B the weight function $w(\theta)$ is chosen for which the maximum of $\beta(\theta)$ with respect to θ (θ restricted to points in ω_r) takes its smallest value.* When this principle for the choice of $w(\theta)$ is adopted, α and the maximum of $\beta(\theta)$ with respect to θ (θ in ω_r) will depend only on the quantities A and B.

[6] Instead of defining p_{1n} by some weighted average of the type given in (4:4), it would seem equally reasonable to define p_{1n} as the maximum of $f(x_1, \theta)\cdots f(x_n, \theta)$ with respect to θ where θ is restricted to points in ω_r. Then the ratio p_{1n}/p_{0n} would coincide with the so-called likelihood ratio introduced by J. Neyman and E. Pearson and widely used in current test procedures. Our reason for preferring weighted averages is that the theory of such tests seems to be considerably simpler. If p_{1n} were defined by the maximum with respect to θ in ω_r, p_{1n} would no longer be a probability distribution.

[7] In fact, with good approximation the following relations hold: $(1 - \bar{\beta})/\alpha = A$ and $\bar{\beta}/(1 - \alpha) = B$ where $\bar{\beta} = \int_{\omega_r} \beta(\theta)w(\theta)\,d\theta$. Solving these equations with respect to α and $\bar{\beta}$ we obtain $\alpha = (1 - B)/(A - B)$ and $\bar{\beta} = [B(A - 1)]/(A - B)$. Thus, α and $\bar{\beta}$ depend only on A and B.

These values A and B are then determined so that the probability of an error of the first kind has the desired value α and the maximum of $\beta(\theta)$ with respect to θ is equal to the required value β.

There is no general method yet available for the determination of an optimum weight function $w(\theta)$ in the sense defined above. For some special but important cases, however, such a weight function has been determined. This point is discussed in Section A.8.

4.1.4 Application of the General Procedure to Testing the Mean of a Normal Distribution with Known Variance

In this section we shall consider the problem of testing the simple hypothesis H_0 that the mean θ of a normal distribution with known variance is equal to a particular value θ_0. The acceptance of H_0 will not be considered a serious error if $\theta \neq \theta_0$ but is near θ_0. However, there will be, in general, a positive value δ such that the acceptance of H_0 is considered an error of practical importance if (and only if) $\left|\dfrac{\theta - \theta_0}{\sigma}\right| \geq \delta$, where σ denotes the known standard deviation of the distribution. Thus, the region of preference for rejection may be defined as the set of all values θ for which $\left|\dfrac{\theta - \theta_0}{\sigma}\right| \geq \delta$. The region of preference for acceptance will consist of the single value θ_0, and the region of indifference will be the set of all values θ for which $0 < \left|\dfrac{\theta - \theta_0}{\sigma}\right| < \delta$.

The probability density of the sample $(x_1, \cdots, x_n)$ under H_0 is given by

$$(4{:}8) \qquad p_{0n} = \frac{1}{(2\pi)^{\frac{n}{2}}\sigma^n}\, e^{-\frac{1}{2\sigma^2}\sum_{\alpha=1}^{n}(x_\alpha-\theta_0)^2}$$

According to the general theory discussed in the preceding section, p_{1n} is defined as some weighted average of the probability density corresponding to various values of θ in the zone of preference for rejection. It is shown in Section A.8.2 that an optimum weighted average is the simple average of the two density functions: the density function corresponding to $\theta = \theta_0 - \delta\sigma$ and the density function corresponding to $\theta = \theta_0 + \delta\sigma$. Thus,

$$(4{:}9) \quad p_{1n} = \frac{1}{2}\left[\frac{1}{(2\pi)^{\frac{n}{2}}\sigma^n}\, e^{-\frac{1}{2\sigma^2}\Sigma(x_\alpha-\theta_0+\delta\sigma)^2} + \frac{1}{(2\pi)^{\frac{n}{2}}\sigma^n}\, e^{-\frac{1}{2\sigma^2}\Sigma(x_\alpha-\theta_0-\delta\sigma)^2}\right]$$

The test is then carried out as follows. We continue taking observations as long as $B < p_{1n}/p_{0n} < A$. If $p_{1n}/p_{0n} \geqq A$, we reject H_0. If $p_{1n}/p_{0n} \leqq B$, we accept H_0. To make the probability of an error of the first kind equal to α and the maximum of $\beta(\theta)$ (in the domain $\left| \dfrac{\theta - \theta_0}{\sigma} \right| \geqq \delta$) equal to β, for all practical purposes we may put $A = (1 - \beta)/\alpha$ and $B = \beta/(1 - \alpha)$.

A more detailed discussion of this test procedure is given in Part II, Chapter 9.

4.2   Tests of Composite Hypotheses

4.2.1   Discussion of an Important Special Case

A frequent and important problem is that of testing the hypothesis H that the unknown parameter θ does not exceed a specified value θ'.[8] This problem is of particular importance in quality control of manufactured products. The importance of an error of the first kind (rejection of H when H is true), or that of an error of the second kind (acceptance of H when H is false), will usually vary with the value of θ. For example, if θ is only slightly below θ' the rejection of H will not be considered a serious error. Similarly, if θ is only slightly above θ' the acceptance of H will not be considered a serious error. In general, the importance of an error of the first kind will increase steadily with decreasing value of θ in the domain $\theta \leqq \theta'$, and the importance of an error of the second kind will increase steadily with increasing value of θ in the domain $\theta > \theta'$. Thus, it will be possible to find two values $\theta_0 < \theta'$ and $\theta_1 > \theta'$ such that an error of the first kind is considered of practical importance whenever $\theta \leqq \theta_0$, and an error of the second kind is considered of practical importance whenever $\theta \geqq \theta_1$, whereas for values θ between θ_0 and θ_1 we do not care particularly which decision is made. Hence the zone of preference for acceptance may be defined as consisting of all values $\theta \leqq \theta_0$, the zone of preference for rejection as the set of values θ for which $\theta \geqq \theta_1$, and the zone of indifference as the set of all values θ for which $\theta_0 < \theta < \theta_1$. In such a situation we shall want a test procedure for which the probability of an error of the first kind is less than or equal to a preassigned α whenever $\theta \leqq \theta_0$, and the probability of an error of the second kind is less than or equal to a preassigned β whenever $\theta \geqq \theta_1$. In most of the important cases occurring in practice, such as when x has a normal, binomial, or Poisson distribution, and so on, the sequential probability

[8] It is assumed here that there is only one unknown parameter θ involved in the distribution of x.

ratio test of strength (α, β) for testing the hypothesis that $\theta = \theta_0$ against the single alternative that $\theta = \theta_1$ will have the desired properties and provides a satisfactory solution to the problem. If the sequential probability ratio test leads to the acceptance of the hypothesis that $\theta = \theta_0$, we accept the original hypothesis that $\theta \leqq \theta'$, and if the probability ratio test leads to the rejection of the hypothesis that $\theta = \theta_0$, we reject the original hypothesis that $\theta \leqq \theta'$.

As an illustration, we shall discuss briefly one or two examples. Suppose that a lot consisting of a large number of units of a manufactured product is submitted for acceptance inspection. We shall assume that each unit is classified in one of the two categories: defective and non-defective. The proportion p of defectives in the lot is assumed to be unknown. The preference for acceptance or rejection of the lot will, of course, depend on the value of p. It will be possible, in general, to select two values of p, say p_0 and p_1 $(p_0 < p_1)$ such that the rejection of the lot is considered an error of practical importance whenever $p \leqq p_0$, and the acceptance of the lot is an error of practical importance whenever $p \geqq p_1$; for values p between p_0 and p_1 we do not care particularly which decision is made. Thus, the zone of preference for acceptance is given by $p \leqq p_0$, the zone of preference for rejection by $p \geqq p_1$, and the zone of indifference consists of values p for which $p_0 < p < p_1$. Hence, we shall want a test procedure for which the probability of rejecting the lot is less than or equal to a preassigned value α whenever $p \leqq p_0$, and the probability of accepting the lot is less than or equal to a preassigned value β whenever $p \geqq p_1$. Such a test procedure is given by the sequential probability ratio test of strength (α, β) for testing the hypothesis that $p = p_0$ against the single alternative that $p = p_1$. To compute the probability ratio p_{1n}/p_{0n} for this problem, we shall denote by d_n the number of defectives found in the first n units inspected. The probability of obtaining a sample equal to the observed one is given by

$$(4{:}10) \qquad\qquad p_{1n} = p_1{}^{d_n}(1 - p_1)^{n-d_n}$$

when $p = p_1$, and by

$$(4{:}11) \qquad\qquad p_{0n} = p_0{}^{d_n}(1 - p_0)^{n-d_n}$$

when $p = p_0$.[9] Then

$$(4{:}12) \qquad \log \frac{p_{1n}}{p_{0n}} = d_n \log \frac{p_1}{p_0} + (n - d_n) \log \frac{1 - p_1}{1 - p_0}$$

[9] Formulas (4:10) and (4:11) are strictly valid only if the lot contains infinitely many units. It is assumed that the lot contains a large number of units so that these formulas can be used with good approximation.

The test procedure is carried out as follows. We continue inspection as long as $\log B < \log (p_{1n}/p_{0n}) < \log A$. If $\log (p_{1n}/p_{0n}) \geqq \log A$, inspection is terminated with the rejection of the lot, and if $\log (p_{1n}/p_{0n}) \leqq \log B$, inspection is terminated with the acceptance of the lot. For practical purposes we may put $A = (1 - \beta)/\alpha$ and $B = \beta/(1 - \alpha)$.

A detailed discussion of the problem of acceptance inspection when each unit is classified either as defective or as non-defective is given in Part II in Chapter 5.

Another example for testing a hypothesis that $\theta \leqq \theta'$ is the case when θ is the unknown mean of a normal distribution with known variance.[10] Again it will be possible to select two values $\theta_0 < \theta'$ and $\theta_1 > \theta'$ such that an error of the first kind is considered of practical importance whenever $\theta \leqq \theta_0$, an error of the second kind is of practical importance whenever $\theta \geqq \theta_1$; for values θ between θ_0 and θ_1 we do not care particularly which decision is made. In such a situation we shall want a test procedure for which the probability of committing an error of the first kind is less than or equal to some preassigned value α whenever $\theta \leqq \theta_0$, and the probability of committing an error of the second kind does not exceed a preassigned value β whenever $\theta \geqq \theta_1$. These conditions will be satisfied by the sequential probability ratio test of strength (α, β) for testing the hypothesis that $\theta = \theta_0$ against the single alternative hypothesis that $\theta = \theta_1$. The probability density of the sample $(x_1, \cdots, x_n)$ is given by

$$(4{:}13) \qquad p_{0n} = \frac{1}{(2\pi)^{\frac{n}{2}}\sigma^n} e^{-\frac{1}{2\sigma^2} \Sigma(x_\alpha - \theta_0)^2}$$

when $\theta = \theta_0$, and by

$$(4{:}14) \qquad p_{1n} = \frac{1}{(2\pi)^{\frac{n}{2}}\sigma^n} e^{-\frac{1}{2\sigma^2} \Sigma(x_\alpha - \theta_1)^2}$$

when $\theta = \theta_1$. We continue taking observations as long as $B < p_{1n}/p_{0n} < A$. If $p_{1n}/p_{0n} \geqq A$, we reject the hypothesis that $\theta \leqq \theta'$, and if $p_{1n}/p_{0n} \leqq B$ we accept the hypothesis that $\theta \leqq \theta'$. Again, we put $A = (1 - \beta)/\alpha$ and $B = \beta/(1 - \alpha)$.

4.2.2 Outline of the Test Procedure in the General Case

In testing a composite hypothesis H_ω that the parameter point θ lies in a subset ω of the parameter space, the parameter space is again subdivided into three mutually exclusive zones: the zone of preference

[10] This problem is discussed in detail in Part II, Chapter 7.

for acceptance ω_a, the zone of preference for rejection ω_r, and the zone of indifference. The zone of preference for acceptance will now consist of more than one parameter point, as distinguished from the case of testing a simple hypothesis.

For any test procedure the probability of an error of the first kind (rejecting H_ω when H_ω is true) will, in general, vary with the parameter point in ω. For any parameter point θ in ω we shall denote by $\alpha(\theta)$ the probability that H_ω will be rejected when θ is true. Similarly, the probability of an error of the second kind (accepting H_ω when it is false) is a function $\beta(\theta)$ defined for all points θ outside ω.

According to the requirements formulated in Section 2.3.2, we shall want a test procedure such that $\alpha(\theta)$ will not exceed a preassigned value α for all θ in the zone ω_a, and $\beta(\theta)$ will not exceed a preassigned value β for all θ in the zone ω_r. Before discussing the problem of finding a proper test procedure satisfying these requirements, we shall again consider, as in the case of the simple hypothesis, the following modified problem: Let $w_a(\theta)$ and $w_r(\theta)$ be two non-negative functions of θ, called weight functions, such that [11]

$$(4\!:\!15) \qquad \int_{\omega_a} w_a(\theta)\, d\theta = 1 \quad \text{and} \quad \int_{\omega_r} w_r(\theta)\, d\theta = 1$$

Suppose that we wish to construct a sequential test such that the weighted average $\int_{\omega_a} \alpha(\theta) w_a(\theta)\, d\theta$ of the probabilities of errors of the first kind is equal to a given value α, and the weighted average $\int_{\omega_r} \beta(\theta) w_r(\theta)\, d\theta$ of the probabilities of errors of the second kind is a given value β.

A proper sequential test satisfying these modified requirements can be constructed as follows. Let p_{0n} and p_{1n} be defined by

$$(4\!:\!16) \qquad p_{0n} = \int_{\omega_a} f(x_1, \theta_1, \cdots, \theta_k) \cdots f(x_n, \theta_1, \cdots, \theta_k) w_a(\theta)\, d\theta$$

and

$$(4\!:\!17) \qquad p_{1n} = \int_{\omega_r} f(x_1, \theta_1, \cdots, \theta_k) \cdots f(x_n, \theta_1, \cdots, \theta_k) w_r(\theta)\, d\theta$$

where $f(x, \theta_1, \cdots, \theta_k)$ denotes the probability distribution of x when θ is true. The functions p_{0n} and p_{1n} can be interpreted as probability distributions of the sample $(x_1, \cdots, x_n)$. Denote by H_0^* the hypoth-

[11] The weight functions $w_a(\theta)$ and $w_r(\theta)$ may also be discrete. Formulas valid for both continuous and discrete weight functions could be given by using Stieltje's integrals in (4:15) and subsequent equations.

esis that the distribution of the sample $(x_1, \cdots, x_n)$ is given by (4:16), and by $H_1{}^*$ the hypothesis that the distribution of $(x_1, \cdots, x_n)$ is given by (4:17). The sequential probability ratio test of strength (α, β) for testing $H_0{}^*$ against $H_1{}^*$ provides a solution to our problem. If the constants A and B in this sequential test are chosen so that the probability is α that we reject $H_0{}^*$ when $H_0{}^*$ is true, and the probability is β that we accept $H_0{}^*$ when $H_1{}^*$ is true, then for this sequential test we have

$$\int_{\omega_a} w_a(\theta)\alpha(\theta)\, d\theta = \alpha$$

and

$$\int_{\omega_r} w_r(\theta)\beta(\theta)\, d\theta = \beta$$

To make the strength of the test of $H_0{}^*$ against $H_1{}^*$ equal to (α, β), again, for practical purposes, we may put $A = (1 - \beta)/\alpha$ and $B = \beta/(1 - \alpha)$.

To construct a sequential test procedure satisfying the requirements

(4:18) $$\alpha(\theta) \leqq \alpha \quad \text{for all } \theta \text{ in } \omega_a$$

and

(4:19) $$\beta(\theta) \leqq \beta \quad \text{for all } \theta \text{ in } \omega_r$$

we shall restrict ourselves to sequential probability ratio tests for which p_{0n} and p_{1n} are given by (4:16) and (4:17), respectively, and $w_a(\theta)$ and $w_r(\theta)$ may be any weight functions satisfying (4:15). Denote by C the class of all such tests corresponding to all possible weight functions $w_a(\theta)$ and $w_r(\theta)$. To select a proper test from the class C which satisfies the requirements (4:18) and (4:19), our procedure will be similar to that in the case of simple hypotheses, as discussed in Section 4.1.3. A test in class C is uniquely determined by the choice of the constants A and B and by the weight functions $w_a(\theta)$ and $w_r(\theta)$. Thus, the maximum of $\alpha(\theta)$ with respect to θ in the zone ω_a, as well as the maximum of $\beta(\theta)$ with respect to θ in the zone ω_r, is determined uniquely by A, B, $w_a(\theta)$, and $w_r(\theta)$. Denote these maxima by $\alpha[A, B, w_a, w_r]$ and $\beta[A, B, w_a, w_r]$, respectively. For given values A and B, the weight functions $w_a(\theta)$ and $w_r(\theta)$ may be regarded the more desirable the smaller they make $\alpha[A, B, w_a, w_r]$ and $\beta[A, B, w_a, w_r]$. Thus, if it is possible to find weight functions $w_a(\theta)$ and $w_r(\theta)$ for which both $\alpha[A, B, w_a, w_r]$ and $\beta[A, B, w_a, w_r]$ are simultaneously minimized, they may be regarded as optimum weight functions. It is shown in Section A.9 that in some important special cases, such as testing the mean of a normal distribution with unknown variance, optimum weight functions of the type described above do exist. However, it is not known whether they generally exist. If it is not possible to minimize both $\alpha[A, B, w_a, w_r]$ and $\beta[A, B, w_a, w_r]$ simultaneously, it may be reasonable to choose $w_a(\theta)$ and $w_r(\theta)$ such that some average of the two values $\alpha[A, B, w_a, w_r]$ and $\beta[A, B, w_a, w_r]$, or the maximum of these two values, is minimized.

If the principle described above for choosing the weight functions $w_a(\theta)$ and $w_r(\theta)$ is adopted, the maximum of $\alpha(\theta)$ in the zone ω_a and the maximum of $\beta(\theta)$ in the zone ω_r will depend only on A and B. Finally the constants A and B are determined so that these two maxima are equal to α and β, respectively.

There is no general method yet available for constructing weight functions $w_a(\theta)$ and $w_r(\theta)$ which are optimum in the sense defined above. In some special cases, however, such weight functions have been constructed.[12]

4.2.3 Application of the General Procedure to Testing the Mean of a Normal Distribution with Unknown Variance (Sequential *t*-Test)

A frequent and important problem in applications is that of testing the hypothesis H that the unknown mean θ of a normal distribution is equal to some specified value θ_0 when nothing is known about the variance σ^2 of the distribution. If the true value θ differs only slightly from θ_0, i.e., if $|\theta - \theta_0|$ is only a small fraction of the standard deviation σ, the acceptance of H will usually not be considered an error of practical consequence. However, the importance of an error committed by accepting H when $\theta \neq \theta_0$ will, in general, increase with increasing value of $\left| \dfrac{\theta - \theta_0}{\sigma} \right|$. Thus, it will be possible to find a positive value δ such that the acceptance of H is considered an error of practical importance only when $\left| \dfrac{\theta - \theta_0}{\sigma} \right| \geq \delta$. Accordingly, the three zones in the parameter space will be defined as follows. The zone ω_a of preference for acceptance consists of all points (θ, σ) for which $\theta = \theta_0$, i.e., ω_a consists of all points (θ_0, σ) where σ can take any positive value. The zone ω_r of preference for rejection consists of all points (θ, σ) for which $\left| \dfrac{\theta - \theta_0}{\sigma} \right| \geq \delta$. Finally the zone of indifference contains all points (θ, σ) for which $0 < \left| \dfrac{\theta - \theta_0}{\sigma} \right| < \delta$.

[12] See Section A.9.

The probability density of a sample $(x_1, \cdots, x_n)$ drawn from a normal distribution with mean θ and standard deviation σ is given by

$$(4\!:\!20) \qquad p_n = \frac{1}{(2\pi)^{\frac{n}{2}}\sigma^n} e^{-\frac{1}{2\sigma^2}\sum\limits_{\alpha=1}^{n}(x_\alpha - \theta)^2}$$

As in the general procedure described in the preceding section, the test procedure will be based on the ratio p_{1n}/p_{0n} where p_{0n} is some weighted average value of p_n corresponding to various points (θ, σ) in ω_a, and p_{1n} is some weighted average of p_n corresponding to various points (θ, σ) in ω_r. It is shown in Section A.9 that by choosing the weight functions $w_a(\theta)$ and $w_r(\theta)$ according to the principles described in the preceding section we are led to the following ratio: [13]

$$(4\!:\!21) \qquad \frac{p_{1n}}{p_{0n}} = \frac{\dfrac{1}{2}\displaystyle\int_0^\infty \frac{1}{\sigma^n}\left[e^{-\frac{1}{2\sigma^2}\sum\limits_{\alpha=1}^{n}(x_\alpha-\theta_0-\delta\sigma)^2} + e^{-\frac{1}{2\sigma^2}\sum\limits_{\alpha=1}^{n}(x_\alpha-\theta_0+\delta\sigma)^2}\right]d\sigma}{\displaystyle\int_0^\infty \frac{1}{\sigma^n} e^{-\frac{1}{2\sigma^2}\Sigma(x_\alpha-\theta_0)^2}\,d\sigma}$$

The test procedure is then carried out as follows. Additional observations are taken as long as $B < p_{1n}/p_{0n} < A$. The hypothesis H is rejected if $p_{1n}/p_{0n} \geqq A$ and the hypothesis H is accepted if $p_{1n}/p_{0n} \leqq B$. To satisfy the requirements (4:18) and (4:19) for practical purposes we may let $A = (1 - \beta)/\alpha$ and $B = \beta/(1 - \alpha)$.

4.2.4   A Particular Class of Problems Treated by Girshick [14]

A class of problems treated by M. A. Girshick may be formulated as follows. Let x_1 and x_2 be two independent random variables. The distribution (elementary probability law) of x_1 is given by $f(x_1, \theta_1)$ and that of x_2 by $f(x_2, \theta_2)$, where the function f is known but the values of the parameters θ_1 and θ_2 are unknown. The problem is to test the hypothesis H that $\theta_1 \leqq \theta_2$ against the alternative hypothesis H' that $\theta_1 > \theta_2$.

The type of problem described above occurs frequently in applications. For example, let x denote some quality characteristic, such as hardness, tensile strength, or weight, of a manufactured product. Sup-

---

[13] Considerable work on the evaluation of this ratio to bring it to a suitable form for tabulation was done by K. Arnold while he was a member of the Statistical Research Group of Columbia University. Tables for the computation of this ratio have been prepared by the Mathematical Tables Project, New York.

[14] M. A. Girshick, "Contributions to the Theory of Sequential Analysis," *The Annals of Mathematical Statistics, Vol.* 17 (1946).

pose that the distribution of x in the population of units produced has a known functional form $f(x, \theta)$, but the value of the parameter θ is unknown. Suppose, furthermore, that there are two competing processes of production under consideration by the manufacturer. Let θ_1 denote the value of θ when process 1 is used, and θ_2 when process 2 is used. Both values, θ_1 and θ_2, are unknown. If the product is considered the more desirable the greater the value of θ, the problem of deciding between the two competing processes reduces to that of testing the hypothesis H that $\theta_1 \leqq \theta_2$. Process 1 is chosen if H is rejected, and process 2 is chosen if H is accepted.

The following procedure for testing the hypothesis H has been proposed by Girshick. We choose a particular value $\theta_1{}^0$ of θ_1 and a particular value $\theta_2{}^0$ of θ_2 where $\theta_1{}^0 < \theta_2{}^0$. Let H_0 denote the hypothesis that the joint distribution of x_1 and x_2 is given by $f(x_1, \theta_1{}^0)f(x_2, \theta_2{}^0)$, and let H_1 be the alternative hypothesis that the joint distribution of x_1 and x_2 is given by $f(x_1, \theta_2{}^0)f(x_2, \theta_1{}^0)$. We then set up the sequential probability ratio test for testing the simple hypothesis H_0 against the simple alternative H_1. The hypothesis H is accepted or rejected accordingly as the sequential probability ratio test leads to the acceptance or rejection of H_0. Thus, to carry out the test procedure, two constants A and B are chosen and the ratio

$$(4\!:\!22) \qquad \frac{p_{1m}}{p_{0m}} = \frac{f(x_{11}, \theta_1{}^0)f(x_{21}, \theta_2{}^0) \cdots f(x_{1m}, \theta_1{}^0)f(x_{2m}, \theta_2{}^0)}{f(x_{11}, \theta_2{}^0)f(x_{21}, \theta_1{}^0) \cdots f(x_{1m}, \theta_2{}^0)f(x_{2m}, \theta_1{}^0)}$$

is computed at each stage of the experiment. Here $x_{i\alpha}$ denotes the αth observation on x_i $(i = 1, 2)$. It is assumed that the observations are taken in pairs, where each pair consists of an observation on x_1 and an observation on x_2. Experimentation is continued as long as the ratio p_{1m}/p_{0m} lies between B and A. The hypothesis H is accepted if $p_{1m}/p_{0m} \leqq B$, and the hypothesis H is rejected if $p_{1m}/p_{0m} \geqq A$.

It has been shown by Girshick that in many important cases the above test procedure will have the following property: There exists a function $v = v(\theta_1, \theta_2)$ such that v may be regarded as a reasonable measure of the difference between θ_1 and θ_2, and the probability of accepting H depends only on the value of v. The function v satisfies, furthermore, the conditions: (1) $v(\theta_1, \theta_2) = 0$ when $\theta_1 = \theta_2$; (2) $v(\theta_1, \theta_2) < 0$ when $\theta_2 > \theta_1$; (3) $v(\theta_1, \theta_2) = -v(\theta_2, \theta_1)$.

If a function v with the above properties exists, the choice of the four quantities $\theta_1{}^0$, $\theta_2{}^0$, A, and B may be made on the basis of the following considerations: Let δ be a positive value such that the acceptance of H is regarded as an error of practical importance whenever

$v \geqq \delta$, the rejection of H is regarded as an error of practical importance whenever $v \leqq -\delta$; for values v between $-\delta$ and δ we do not care particularly which decision is made. Thus, we shall want a test procedure for which the probability of rejecting H will not exceed a preassigned value α whenever $v \leqq -\delta$ and the probability of accepting H will not exceed a preassigned value β whenever $v \geqq \delta$. The test procedure will have the desired properties if the quantities $\theta_1{}^0$, $\theta_2{}^0$, A, and B are chosen so that $v(\theta_1{}^0, \theta_2{}^0) = -\delta$ and the sequential probability ratio test for testing H_0 against H_1 has the strength (α, β). For all practical purposes we may let $A = (1 - \beta)/\alpha$ and $B = \beta/(1 - \alpha)$.

As an illustration, we shall consider the following example. Suppose that one of two production processes is to be chosen. Suppose, further, that the quality characteristic under consideration is normally distributed with known mean and unknown standard deviation σ_1 when process 1 is used, and that the distribution is normal with the same mean but unknown standard deviation σ_2 when process 2 is used. The process that leads to a smaller standard-deviation is preferred. Thus, the manufacturer is interested in testing the hypothesis H that $\sigma_1 \leqq \sigma_2$. There is no loss of generality in assuming that the known means are equal to 0. Let H_0 be the hypothesis that $\sigma_1 = \sigma_1{}^0$ and $\sigma_2 = \sigma_2{}^0$, and H_1 the hypothesis that $\sigma_1 = \sigma_2{}^0$ and $\sigma_2 = \sigma_1{}^0$ $(\sigma_1{}^0 < \sigma_2{}^0)$. Then the probability ratio for testing H_0 against H_1 is given by

$$(4{:}23) \qquad \frac{p_{1m}}{p_{0m}} = e^{\left(\frac{1}{2(\sigma_1{}^0)^2} - \frac{1}{2(\sigma_2{}^0)^2}\right)\left[\sum\limits_{\alpha=1}^{m} (x_{1\alpha}{}^2 - x_{2\alpha}{}^2)\right]}$$

where $x_{i\alpha}$ denotes the αth observation from the population corresponding to process i.

As Girshick has shown, the probability that the sequential probability ratio test of H_0 against H_1 will terminate with the acceptance of H_0 depends only on the value of

$$(4{:}24) \qquad v(\sigma_1, \sigma_2) = \frac{1}{2}\left(\frac{1}{\sigma_2{}^2} - \frac{1}{\sigma_1{}^2}\right)$$

This quantity may be regarded as a reasonable measure of the deviation of σ_1 from σ_2. Suppose we want a test procedure satisfying the following conditions: The probability of rejecting H should not exceed α whenever $\frac{1}{2}\left(\frac{1}{\sigma_2{}^2} - \frac{1}{\sigma_1{}^2}\right) \leqq -\delta$, and the probability of accepting H should not exceed β whenever $\frac{1}{2}\left(\frac{1}{\sigma_2{}^2} - \frac{1}{\sigma_1{}^2}\right) \geqq \delta$. Then we choose

$\sigma_1{}^0$ and $\sigma_2{}^0$ so that

(4:25) $$\frac{1}{2}\left(\frac{1}{(\sigma_2{}^0)^2} - \frac{1}{(\sigma_1{}^0)^2}\right) = -\delta$$

The probability ratio given in (4:23) becomes then equal to

(4:26) $$\frac{p_{1m}}{p_{0m}} = e^{\delta \sum\limits_{\alpha=1}^{m} (x_{1\alpha}{}^2 - x_{2\alpha}{}^2)}$$

When $\dfrac{1}{\delta} \log \dfrac{p_{1m}}{p_{0m}} = \Sigma(x_{1\alpha}{}^2 - x_{2\alpha}{}^2)$ is used instead of $\dfrac{p_{1m}}{p_{0m}}$, the test procedure can be carried out as follows. We continue taking pairs of observations as long as

(4:27) $$\frac{\log B}{\delta} < \sum_{\alpha=1}^{m} (x_{1\alpha}{}^2 - x_{2\alpha}{}^2) < \frac{\log A}{\delta}$$

We accept H if

(4:28) $$\sum_{\alpha=1}^{m}(x_{1\alpha}{}^2 - x_{2\alpha}{}^2) \leqq \frac{\log B}{\delta}$$

and reject H if

(4:29) $$\sum_{\alpha=1}^{m}(x_{1\alpha}{}^2 - x_{2\alpha}{}^2) \geqq \frac{\log A}{\delta}$$

PART II. APPLICATION OF THE GENERAL THEORY TO SPECIAL CASES [1]

Chapter 5. TESTING THE MEAN OF A BINOMIAL DISTRIBUTION (ACCEPTANCE INSPECTION OF A LOT WHERE EACH UNIT IS CLASSIFIED INTO ONE OF TWO CATEGORIES)

5.1 Formulation of the Problem

Let x be a random variable which can take only the values 0 and 1. Denote by p the (unknown) probability that x takes the value 1. We shall deal here with the problem of testing the hypothesis that p does not exceed some specified value p'.

This problem arises, for example, in acceptance inspection of a lot consisting of a large number of units of a manufactured product. Suppose that each unit is classified in one of the two categories: defective and non-defective. We shall assign the value 0 to any non-defective unit and the value 1 to any defective unit. Let p denote the unknown proportion of defectives in the lot. Then the result x of the inspection of a unit drawn at random from the lot can take only the values 1 and 0 with probabilities p and $1 - p$, respectively. Usually it will be possible to specify some value p' such that we would like to accept the lot whenever $p \leq p'$ and we would like to reject the lot whenever $p > p'$. Thus, the problem of deciding whether the lot is to be accepted or rejected on the basis of a random sample may be formulated as the problem of testing the hypothesis $p \leq p'$ against the alternative hypothesis that $p > p'$.

Since acceptance inspection of manufactured products is perhaps one of the most important applications of testing the mean of a binomial distribution, in what follows we shall use the terminology cus-

---

[1] The special cases treated here are discussed mainly to illustrate the general theory and to bring out points of theoretical interest specific to these applications. Accordingly, computational procedures and simplifications are not stressed much and hardly any tables are given. A more detailed and non-mathematical discussion of these applications, together with a number of tables, charts, and computational simplifications, is contained in "Sequential Analysis of Statistical Data: Applications," a report prepared by the Statistical Research Group of Columbia University and published by Columbia University Press, Sept., 1945. This report will be referred to hereafter simply as SRG 255.

tomary in acceptance inspection. This, of course, does not mean that the test procedure is not applicable to other cases as well. In the terminology of acceptance inspection, our problem may be stated as follows: A proper sampling plan (test procedure) is to be devised for deciding whether the lot submitted for inspection should be accepted or rejected.

5.2   Tolerated Risks of Making Wrong Decisions

Any sampling plan which does not provide for complete inspection of the lot may lead to a wrong decision. That is, we may accept the lot when $p > p'$, or we may reject the lot when $p \leq p'$. Since complete inspection is frequently not feasible, or too costly, we are willing to tolerate some risks of making wrong decisions. In order to devise a proper sampling plan, it is necessary to state the maximum risks of wrong decisions that we are willing to tolerate.

If $p = p'$, the quality of the lot is just on the margin and we are indifferent which decision is made. For $p > p'$, we prefer to reject the lot and this preference increases with increasing value of p. For $p < p'$, we prefer to accept the lot and this preference increases with decreasing value of p. If p is only slightly above p', the preference for rejection is only slight and acceptance of the lot will not be regarded as an error of practical consequence. Similarly, if p is only slightly below p', rejection of the lot is not a serious error. Thus, it will be possible to specify two values p_0 and p_1, p_0 below p' and p_1 above p', such that acceptance of the lot is regarded as an error of practical consequence if (and only if) $p \geq p_1$, and rejection of the lot is regarded as an error of practical importance if (and only if) $p \leq p_0$. If p lies between p_0 and p_1 we do not care particularly which decision is made.

After the two values p_0 and p_1 have been chosen, the risks of making wrong decisions which we are willing to tolerate may reasonably be formulated as follows: The probability of rejecting the lot should not exceed some small preassigned value α whenever $p \leq p_0$, and the probability of accepting the lot should not exceed some small preassigned value β whenever $p \geq p_1$.

Thus, the tolerated risks are characterized by four numbers, p_0, p_1, α, and β. The choice of these four quantities is not a statistical problem. They will be selected on the basis of practical considerations in each particular case. A proper sampling plan can be determined, as will be shown in the next section, after these four quantities have been chosen.

5.3   The Sequential Probability Ratio Test Corresponding to the Quantities p_0, p_1, α, and β

5.3.1   Derivation of Algebraic Formulas for the Test Criterion

A sampling plan satisfying the conditions that the probability of rejecting the lot does not exceed α whenever $p \leqq p_0$, and the probability of accepting the lot does not exceed β whenever $p \geqq p_1$, is given by the sequential probability ratio test of strength (α, β) for testing the hypothesis $p = p_0$ against the hypothesis $p = p_1$. This test is defined as follows (see Section 3.1): Let x_i denote the result of the inspection of the ith unit; that is, $x_i = 1$ if the ith unit inspected is found defective, and $x_i = 0$ otherwise. If p denotes the proportion of defectives in the lot, the probability of obtaining a sample equal to the observed $(x_1, \cdots, x_m)$ is given by

$$(5\!:\!1) \qquad\qquad p^{d_m}(1 - p)^{m - d_m}$$

where d_m denotes the number of defectives in the first m units inspected.[2] Under the hypothesis that $p = p_1$ the probability (5:1) becomes equal to

$$(5\!:\!2) \qquad\qquad p_{1m} = p_1{}^{d_m}(1 - p_1)^{m - d_m}$$

and under the hypothesis that $p = p_0$ the probability (5:1) becomes equal to

$$(5\!:\!3) \qquad\qquad p_{0m} = p_0{}^{d_m}(1 - p_0)^{m - d_m}$$

The sequential probability ratio test is carried out as follows. At each stage of the inspection, at the inspection of the mth unit for each positive integral value m, we compute

$$(5\!:\!4) \qquad \log \frac{p_{1m}}{p_{0m}} = d_m \log \frac{p_1}{p_0} + (m - d_m) \log \frac{1 - p_1}{1 - p_0}$$

Inspection is continued as long as

$$(5\!:\!5) \qquad\qquad \log \frac{\beta}{1 - \alpha} < \log \frac{p_{1m}}{p_{0m}} < \log \frac{1 - \beta}{\alpha}$$

[2] The lot is assumed to be sufficiently large so that the successive observations x_1, x_2, $\cdots$, etc., may be regarded as independent.

Inspection is terminated the first time that (5:5) does not hold. If at this final stage we have

$$(5:6) \qquad \log \frac{p_{1m}}{p_{0m}} \geqq \log \frac{1-\beta}{\alpha}$$

the lot is rejected, and if

$$(5:7) \qquad \log \frac{p_{1m}}{p_{0m}} \leqq \log \frac{\beta}{1-\alpha}$$

the lot is accepted.[3]

Inequalities (5:5), (5:6), and (5:7) can easily be seen to be equivalent to the following inequalities:

$$(5:8) \qquad \frac{\log \dfrac{\beta}{1-\alpha}}{\log \dfrac{p_1}{p_0} - \log \dfrac{1-p_1}{1-p_0}} + m \frac{\log \dfrac{1-p_0}{1-p_1}}{\log \dfrac{p_1}{p_0} - \log \dfrac{1-p_1}{1-p_0}} < d_m <$$

$$\frac{\log \dfrac{1-\beta}{\alpha}}{\log \dfrac{p_1}{p_0} - \log \dfrac{1-p_1}{1-p_0}} + m \frac{\log \dfrac{1-p_0}{1-p_1}}{\log \dfrac{p_1}{p_0} - \log \dfrac{1-p_1}{1-p_0}}$$

$$(5:9) \qquad d_m \geqq \frac{\log \dfrac{1-\beta}{\alpha}}{\log \dfrac{p_1}{p_0} - \log \dfrac{1-p_1}{1-p_0}} + m \frac{\log \dfrac{1-p_0}{1-p_1}}{\log \dfrac{p_1}{p_0} - \log \dfrac{1-p_1}{1-p_0}}$$

and

$$(5:10) \qquad d_m \leqq \frac{\log \dfrac{\beta}{1-\alpha}}{\log \dfrac{p_1}{p_0} - \log \dfrac{1-p_1}{1-p_0}} + m \frac{\log \dfrac{1-p_0}{1-p_1}}{\log \dfrac{p_1}{p_0} - \log \dfrac{1-p_1}{1-p_0}}$$

For each value of m we shall denote the right-hand member of (5:10) by a_m and call it acceptance number. Similarly, we shall denote the

[3] There is a slight approximation involved in the use of the constants $\log [\beta/(1-\alpha)]$ and $\log [(1-\beta)/\alpha]$. For further details see Section 3.3.

right-hand member of (5:9) by r_m and call it rejection number. For purposes of practical computations, the use of the inequalities (5:8), (5:9), and (5:10) seems to be much more convenient than the use of the original inequalities (5:5), (5:6), and (5:7).[4] On the basis of inequalities (5:8), (5:9), and (5:10), the sequential probability ratio test is carried out as follows. At each stage of the inspection we compute the acceptance number a_m and the rejection number r_m. Inspection is continued as long as $a_m < d_m < r_m$. The first time that d_m does not lie between the acceptance and rejection numbers, inspection is terminated. If $d_m \geqq r_m$ the lot is rejected, and if $d_m \leqq a_m$ the lot is accepted.

5.3.2 Tabular Procedure for Carrying Out the Test

The acceptance number

$$(5:11) \qquad a_m = \frac{\log \dfrac{\beta}{1-\alpha}}{\log \dfrac{p_1}{p_0} - \log \dfrac{1-p_1}{1-p_0}} + m \frac{\log \dfrac{1-p_0}{1-p_1}}{\log \dfrac{p_1}{p_0} - \log \dfrac{1-p_1}{1-p_0}}$$

and the rejection number

$$(5:12) \qquad r_m = \frac{\log \dfrac{1-\beta}{\alpha}}{\log \dfrac{p_1}{p_0} - \log \dfrac{1-p_1}{1-p_0}} + m \frac{\log \dfrac{1-p_0}{1-p_1}}{\log \dfrac{p_1}{p_0} - \log \dfrac{1-p_1}{1-p_0}}$$

depend only on the quantities p_0, p_1, α, and β. Thus, they can be computed and tabulated before inspection starts. If a_m is not an integer, we may replace it by the largest integer $< a_m$. Similarly, if r_m is not an integer, we may replace it by the smallest integer $> r_m$.

As an illustration, consider the following example. Let $p_0 = .1$, $p_1 = .3$, $\alpha = .02$, and $\beta = .03$. The acceptance and rejection numbers, as well as the results of the observations, in an experiment are

---

[4] The use of the inequalities (5:8), (5:9), and (5:10) instead of (5:5), (5:6), and (5:7) was first suggested by J. H. Curtiss. In SRG 255 similar transformations of the inequalities defining the test procedure have been used in other problems as well.

given in Table 5. In this example, inspection is terminated at $m =$ 22 with the rejection of the lot.

TABLE 5

m Number of Units Inspected	a_m Acceptance Number	d_m Number of Defects Observed	r_m Rejection Number
1	..	0	..
2	..	0	..
3	..	1	..
4	..	1	4
5	..	1	4
6	..	1	4
7	..	1	5
8	..	1	5
9	..	2	5
10	..	2	5
11	..	3	5
12	..	4	6
13	..	4	6
14	0	5	6
15	0	5	6
16	0	5	6
17	0	5	7
18	0	6	7
19	0	6	7
20	1	6	7
21	1	6	7
22	1	7	7
23	1	..	8
24	1	..	8
25	2	..	8
26	2	..	8
27	2	..	8
28	2	..	9
29	2	..	9
30	3	..	9

5.3.3 Graphical Procedure for Carrying Out the Test

The test procedure can also be carried out graphically. The number m of observations is measured along the horizontal axis and the number d_m of defects along the vertical axis. The points (m, a_m) lie on a straight line L_0, since a_m is a linear function of m. Similarly the points (m, r_m) lie on a straight line L_1. The intercept of L_0 is given by

$$(5:13) \qquad h_0 = \frac{\log \dfrac{\beta}{1 - \alpha}}{\log \dfrac{p_1}{p_0} - \log \dfrac{1 - p_1}{1 - p_0}}$$

and the intercept of L_1 is given by

$$(5:14) \qquad h_1 = \frac{\log \dfrac{1 - \beta}{\alpha}}{\log \dfrac{p_1}{p_0} - \log \dfrac{1 - p_1}{1 - p_0}}$$

The lines L_0 and L_1 are parallel and the common slope is equal to

$$(5:15) \qquad s = \frac{\log \dfrac{1 - p_0}{1 - p_1}}{\log \dfrac{p_1}{p_0} - \log \dfrac{1 - p_1}{1 - p_0}}$$

The two straight lines L_0 and L_1 are drawn before inspection starts. The points (m, d_m) are plotted as inspection goes on. We continue inspecting additional units as long as the point (m, d_m) lies between the lines L_0 and L_1. Inspection is terminated the first time that the point (m, d_m) does not lie between the lines L_0 and L_1. If (m, d_m) lies on L_0 or below, the lot is accepted. If (m, d_m) lies on L_1 or above, the lot is rejected.

Figure 11 shows the graphical procedure for the example given in Section 5.3.2.

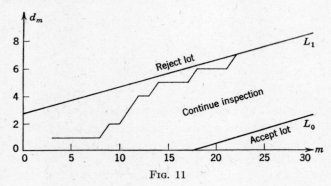

FIG. 11

5.4   The Operating Characteristic (OC) Function $L(p)$ of the Test [5]

5.4.1   Determination of $L(p)$ for Some Special Values of p

As defined in Section 2.2.1, the value of the OC function $L(p)$ for each p is equal to the probability that the lot will be accepted when p is the true proportion of defectives in the lot. One can easily verify that

$$(5{:}16) \qquad L(0) = 1 \quad \text{and} \quad L(1) = 0$$

Since the test procedure is so set up that the probability is $1 - \alpha$ that the lot will be accepted when $p = p_0$, and the probability is β that the lot will be accepted when $p = p_1$, we have

$$(5{:}17) \qquad L(p_0) = 1 - \alpha \quad \text{and} \quad L(p_1) = \beta$$

When
$$p = s = \cfrac{\log \cfrac{1 - p_0}{1 - p_1}}{\log \cfrac{p_1}{p_0} - \log \cfrac{1 - p_1}{1 - p_0}}$$

we obtain from equation (3:43)

$$(5{:}18) \qquad L(s) = \cfrac{\log \cfrac{1 - \beta}{\alpha}}{\log \cfrac{1 - \beta}{\alpha} + \left| \log \cfrac{\beta}{1 - \alpha} \right|} = \frac{h_1}{h_1 + |h_0|}$$

where h_0 and h_1 are the intercepts of the lines L_0 and L_1.[6]

Thus, five points on the OC curve corresponding to $p = 0, 1, p_0, p_1$, and s can immediately be determined. Since $L(p)$ is monotonically decreasing with increasing p, the five points will determine fairly closely the shape of the whole OC curve. This will frequently be sufficient for practical purposes and there will be no need to compute $L(p)$ for additional values of p.

---

[5] The formulas given in this section involve an approximation caused by neglecting the excess of d_m over the boundaries a_m and r_m at the termination of the test procedure. For details see Sections 3.4 and A.2.3. An exact formula for $L(p)$ is given in Section 5.4.3 for the special case in which the slope s of the decision lines is equal to the reciprocal of an integer.

[6] When $p = s$, the value of h in formula (3:43) is equal to 0. The limiting value of the right-hand member of (3:43), when $h \to 0$, is equal to $\dfrac{\log A}{\log A + |\log B|}$ which is equal to the right-hand member of (5:18), since $A = (1 - \beta)/\alpha$ and $B = \beta/(1 - \alpha)$.

5.4.2   Determination of $L(p)$ over the Whole Range of p

It has been shown in Chapter 3, equations (3:45) and (3:46), that [7]

$$(5:19) \qquad L(p) = \frac{\left(\dfrac{1-\beta}{\alpha}\right)^h - 1}{\left(\dfrac{1-\beta}{\alpha}\right)^h - \left(\dfrac{\beta}{1-\alpha}\right)^h}$$

where h is determined by the equation

$$(5:20) \qquad p = \frac{1 - \left(\dfrac{1-p_1}{1-p_0}\right)^h}{\left(\dfrac{p_1}{p_0}\right)^h - \left(\dfrac{1-p_1}{1-p_0}\right)^h}$$

To compute the OC curve, it is not necessary to solve equation (5:20) in h. For any arbitrarily chosen value h, the values of p and $L(p)$ may be computed from (5:19) and (5:20). The point $[p, L(p)]$ computed in this way will be a point on the OC curve. The OC curve can be drawn by plotting a sufficiently large number of points $[p, L(p)]$ corresponding to various values of h. Figure 12 shows a typical OC curve.

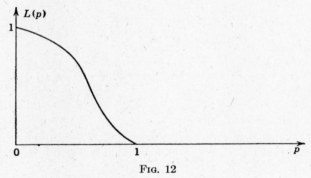

Fᴵɢ. 12

The range of h in (5:19) and (5:20) is from $-\infty$ to $+\infty$. It can be verified that the right-hand member of (5:19) is increasing with increasing h, and the right-hand member of (5:20) is decreasing with increasing h. The five values of p considered in Section 5.4.1, that is, $p = 0, p_0, s, p_1, 1$, correspond to the values of $h = +\infty, 1, 0, -1, -\infty$, respectively, as can be seen from (5:20). Letting $h = +\infty, 1, 0, -1$,

---

[7] In the formulas given in SRG 255, p. 2.50, the quantities p and $L(p)$ are expressed in terms of another parameter x which is functionally related to h.

$-\infty$ in (5:19), we obtain the corresponding five values of $L(p)$ which coincide with those given in Section 5.4.1.

If the part of the OC curve corresponding to positive values of h has been determined, the computation of the part of the OC curve corresponding to negative values of h can be simplified.[8] To show this, let h be a given positive value and let $[p, L(p)]$ be the corresponding point on the OC curve. Let $[p', L(p')]$ denote the point on the OC curve corresponding to $-h$. Then we have

$$(5:21) \quad L(p') = \frac{\left(\dfrac{1-\beta}{\alpha}\right)^{-h} - 1}{\left(\dfrac{1-\beta}{\alpha}\right)^{-h} - \left(\dfrac{\beta}{1-\alpha}\right)^{-h}}$$

$$= \frac{\left(\dfrac{1-\beta}{\alpha}\right)^{h}\left(\dfrac{\beta}{1-\alpha}\right)^{h}\left[\left(\dfrac{1-\beta}{\alpha}\right)^{-h} - 1\right]}{\left(\dfrac{1-\beta}{\alpha}\right)^{h}\left(\dfrac{\beta}{1-\alpha}\right)^{h}\left[\left(\dfrac{1-\beta}{\alpha}\right)^{-h} - \left(\dfrac{\beta}{1-\alpha}\right)^{-h}\right]}$$

$$= \frac{\left(\dfrac{\beta}{1-\alpha}\right)^{h} - \left(\dfrac{1-\beta}{\alpha}\right)^{h}\left(\dfrac{\beta}{1-\alpha}\right)^{h}}{\left(\dfrac{\beta}{1-\alpha}\right)^{h} - \left(\dfrac{1-\beta}{\alpha}\right)^{h}}$$

$$= \left(\dfrac{\beta}{1-\alpha}\right)^{h} \frac{1 - \left(\dfrac{1-\beta}{\alpha}\right)^{h}}{\left(\dfrac{\beta}{1-\alpha}\right)^{h} - \left(\dfrac{1-\beta}{\alpha}\right)^{h}} = \left(\dfrac{\beta}{1-\alpha}\right)^{h} L(p)$$

Similarly,

$$(5:22) \quad p' = \frac{1 - \left(\dfrac{1-p_1}{1-p_0}\right)^{-h}}{\left(\dfrac{p_1}{p_0}\right)^{-h} - \left(\dfrac{1-p_1}{1-p_0}\right)^{-h}} = \frac{\left(\dfrac{p_1}{p_0}\right)^{h}\left(\dfrac{1-p_1}{1-p_0}\right)^{h} - \left(\dfrac{p_1}{p_0}\right)^{h}}{\left(\dfrac{1-p_1}{1-p_0}\right)^{h} - \left(\dfrac{p_1}{p_0}\right)^{h}}$$

$$= \left(\dfrac{p_1}{p_0}\right)^{h} \frac{\left(\dfrac{1-p_1}{1-p_0}\right)^{h} - 1}{\left(\dfrac{1-p_1}{1-p_0}\right)^{h} - \left(\dfrac{p_1}{p_0}\right)^{h}} = \left(\dfrac{p_1}{p_0}\right)^{h} p$$

[8] A similar simplification is given in SRG 255, p. 2.50, with reference to the parameter x used there.

Thus, the point $[p', L(p')]$ corresponding to $-h$ can be computed from the point $[p, L(p)]$ corresponding to h by using the simple relations

$$p' = \left(\frac{p_1}{p_0}\right)^h p \text{ and } L(p') = \left(\frac{\beta}{1-\alpha}\right)^h L(p).$$

5.4.3   Exact Formula for $L(p)$ When the Reciprocal of the Slope of the Decision Lines Is an Integer

The quantity z, i.e., the logarithm of the probability ratio for a single observation, can take only the values $\log(p_1/p_0)$ and $\log[(1-p_1)/(1-p_0)]$. It follows from (5:15) that

$$\log\frac{p_1}{p_0} = \left(\frac{1}{s} - 1\right)\log\frac{1-p_0}{1-p_1}$$

where s is the slope of the decision lines. Assume that $1/s$ is an integer. Then the two values of z are integral multiples of $d = \log[(1-p_0)/(1-p_1)]$, namely, $-d$ and $[(1/s)-1]d$, and the results in the last part of Section A.4 can be used to determine the exact OC curve.[9] On the basis of these results one can show that

$$L(p) = \frac{\displaystyle\sum_{i=1}^{\frac{1}{s}} \frac{u_i^{\frac{1}{s}-2+\left[\frac{\log A}{d}\right]}}{(u_i - 1)\displaystyle\prod_{j\neq i}(u_i - u_j)}}{\displaystyle\sum_{i=1}^{\frac{1}{s}} \frac{u_i^{\frac{1}{s}-2+\left[\frac{\log A}{d}\right]+\left[\frac{|\log B|}{d}\right]}}{(u_i - 1)\displaystyle\prod_{j\neq i}(u_i - u_j)}}$$

where A and B are the constants used in the sequential test,[10] the symbol $[k]$ denotes the smallest integer $\geq k$, and $u_1, u_2, \cdots, u_{\frac{1}{s}}$ are the roots of the equation

$$(1-p)u + p\frac{1}{u^{\frac{1}{\frac{1}{s}-1}}} = 1$$

A different method for deriving an exact formula for $L(p)$ was given by M. A. Girshick in *The Annals of Mathematical Statistics*, Vol. 17 (1946). His method does not require the computation of the roots $u_1, \cdots, u_{\frac{1}{s}}$.

[9] To reduce this case to the case discussed in the last part of Section A.4, one merely has to consider the test corresponding to z^*, A^*, and B^* where $z^* = -z$, $\log A^* = -\log B$ and $\log B^* = -\log A$.

[10] To obtain a test of strength (α, β), we used the approximate values $A = (1-\beta)/\alpha$ and $B = \beta/(1-\alpha)$.

5.5   The Average Sample Number (ASN) Function of the Test

Let n denote the number of observations required by the test procedure. Then n is a random variable, since it depends on the outcome of the observations. The expected value of n depends on the proportion of defectives in the lot and is denoted by $E_p(n)$. This can be plotted as a curve, p being measured along the horizontal axis and $E_p(n)$ along the vertical axis. A typical ASN curve is shown in Fig. 13. This curve is called the ASN curve of the test (see Section 2.2.2 for a general definition of the ASN curve).

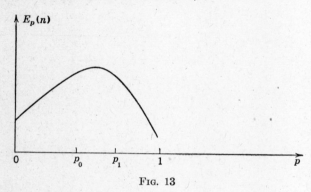

FIG. 13

The general formula for the ASN function of a sequential probability ratio test is derived in Section 3.5. The approximation formula (3:57) applied to the binomial case gives [11]

$$(5:23) \qquad E_p(n) = \frac{L(p) \log B + (1 - L(p)) \log A}{p \log \dfrac{p_1}{p_0} + (1 - p) \log \dfrac{1 - p_1}{1 - p_0}}$$

where $A = (1 - \beta)/\alpha$, $B = \beta/(1 - \alpha)$, and $L(p)$ denotes the probability that inspection terminates with the acceptance of the lot. Using this formula, we shall compute $E_p(n)$ for $p = 0$, p_0, p_1, and 1. Since $L(0) = 1$, the value of $E_p(n)$ is given by

$$(5:24) \qquad E_p(n) = \frac{\log \dfrac{\beta}{1 - \alpha}}{\log \dfrac{1 - p_1}{1 - p_0}}$$

[11] The right-hand member of (5:23) can be expressed as a function of $L(p)$, the intercepts, and the slope of the decision lines. See SRG 255, p. 2.63.

when $p = 0$. For $p = p_0$, we have $L(p) = 1 - \alpha$ and we obtain from (5:23)

$$
(5{:}25) \qquad E_{p_0}(n) = \frac{(1 - \alpha) \log \dfrac{\beta}{1 - \alpha} + \alpha \log \dfrac{1 - \beta}{\alpha}}{p_0 \log \dfrac{p_1}{p_0} + (1 - p_0) \log \dfrac{1 - p_1}{1 - p_0}}
$$

For $p = p_1$, we have $L(p) = \beta$ and we obtain from (5:23)

$$
(5{:}26) \qquad E_{p_1}(n) = \frac{\beta \log \dfrac{\beta}{1 - \alpha} + (1 - \beta) \log \dfrac{1 - \beta}{\alpha}}{p_1 \log \dfrac{p_1}{p_0} + (1 - p_1) \log \dfrac{1 - p_1}{1 - p_0}}
$$

Since $L(1) = 0$, we obtain from (5:23)

$$
(5{:}27) \qquad E_p(n) = \frac{\log \dfrac{1 - \beta}{\alpha}}{\log \dfrac{p_1}{p_0}}
$$

when $p = 1$.

Using formula (A:99) in the Appendix, we can compute the value of $E_p(n)$ when p is equal to the common slope s of the acceptance and rejection lines, i.e., when [12]

$$
p = \frac{\log \dfrac{1 - p_0}{1 - p_1}}{\log \dfrac{p_1}{p_0} - \log \dfrac{1 - p_1}{1 - p_0}} = s
$$

From (A:99) we obtain

$$
(5{:}28) \qquad E_s(n) = \frac{-\left(\log \dfrac{\beta}{1 - \alpha}\right)\left(\log \dfrac{1 - \beta}{\alpha}\right)}{E_s(z^2)}
$$

where $E_s(z^2)$ is the expected value of z^2 and z is a random variable which can take only the values $\log (p_1/p_0)$ and $\log [(1 - p_1)/(1 - p_0)]$

[12] The value s of p corresponds to the value θ' in formula (A:99). It can be shown that s lies between p_0 and p_1. Formula (A:99), and therefore also (5:28), involves an approximation caused by neglecting the excess of the cumulative sum over the boundaries.

with probabilities s and $1 - s$, respectively. Thus

$$(5{:}29) \quad E_s(z^2) = s\left(\log \frac{p_1}{p_0}\right)^2 + (1-s)\left(\log \frac{1-p_1}{1-p_0}\right)^2$$

$$= s\left[\left(\log \frac{p_1}{p_0}\right)^2 - \left(\log \frac{1-p_1}{1-p_0}\right)^2\right] + \left(\log \frac{1-p_1}{1-p_0}\right)^2$$

$$= \left(\log \frac{1-p_0}{1-p_1}\right)\left(\log \frac{p_1}{p_0} + \log \frac{1-p_1}{1-p_0}\right) + \left(\log \frac{1-p_1}{1-p_0}\right)^2$$

$$= \log \frac{p_1}{p_0} \log \frac{1-p_0}{1-p_1}$$

From (5:28) and (5:29) we obtain

$$(5{:}30) \qquad E_s(n) = \frac{-\left(\log \frac{\beta}{1-\alpha}\right)\left(\log \frac{1-\beta}{\alpha}\right)}{\log \frac{p_1}{p_0} \log \frac{1-p_0}{1-p_1}}$$

The determination of the five points of the OC curve, as given in (5:24), (5:25), (5:26), (5:27), and (5:30), may frequently suffice in practice, since these five points already give a fairly good idea of the shape of the whole curve. The ASN curve generally increases as p increases from 0 to p_0, and decreases as p increases from p_1 to 1. In the interval (p_0, p_1) the ASN curve generally increases as p increases from p_0 to some value p', and decreases as p increases from p' to p_1. The value p' is generally equal to s or is very near s.

If it is desired to plot the ASN curve over the whole range of p, it is necessary first to compute the OC function $L(p)$. The value of $E_p(n)$ can then easily be determined from (5:23) for any value p.

5.6   Observations Taken in Groups

5.6.1   General Discussion

For practical reasons it may sometimes be preferable to take the observations in groups, rather than singly. That is, the test procedure is carried out as follows. A group g_1 consisting of v units is drawn from the lot. If the number of defectives d_v in this group g_1 is less than or equal to the acceptance number a_v, inspection terminates with the acceptance of the lot. If d_v is greater than or equal to the rejection number r_v, inspection terminates with the rejection of the lot. If

$a_v < d_v < r_v$, a second group g_2 of v units is drawn. Again, the lot is accepted if the total number of defectives d_{2v} in the two groups is less than or equal to a_{2v}, the lot is rejected if $d_{2v} \geqq r_{2v}$, and a third group g_3 of v units is drawn if $a_{2v} < d_{2v} < r_{2v}$. This process is continued until either rejection or acceptance of the lot is decided. Thus, when the observations are taken in groups of v units, the number d_m of defectives found is compared with the corresponding acceptance number a_m and rejection number r_m only for $m = v, 2v, 3v, \cdots$, etc.

The purpose of this section is to make some comments on the effect of grouping on the OC and ASN curves of the sampling plan. Clearly, grouping can only increase the number of observations required by the test. For, suppose that inspection terminates at the nth unit when observations are taken singly. If n is equal to an integral multiple of v, i.e., $n = kv$, then the number of groups inspected, when observations are taken in groups, will be precisely equal to k, and the total number of units inspected will be the same as when observations are taken singly. However, if $kv < n < (k+1)v$, grouping will cause an increase in the amount of inspection, since we shall have to inspect at least $(k+1)$ groups, that is, at least $(k+1)v$ units. It may even happen that we shall have to inspect more than $(k+1)$ groups. This will be the case when d_n lies outside the interval (a_n, r_n), but $a_{(k+1)v} < d_{(k+1)v} < r_{(k+1)v}$. Thus, the increase in the expected number of units inspected caused by grouping may even exceed v in some cases.

Regarding the effect of grouping on the OC curve, the following remarks may be made. Putting $A = (1 - \beta)/\alpha$ and $B = \beta/(1 - \alpha)$, the probability α' of rejecting the lot when $p = p_0$ and the probability β' of accepting the lot when $p = p_1$ will be only approximately equal to α and β, respectively, even if the observations are taken singly. This was pointed out in Section 3.3, where the following inequalities were derived:

$$(5{:}31) \qquad \alpha' \leqq \frac{\alpha}{1 - \beta} \quad \text{and} \quad \beta' \leqq \frac{\beta}{1 - \alpha}$$

It can easily be verified that these inequalities also remain valid when the observations are taken in groups. The quantities α and β are usually very small and $\alpha/(1 - \beta)$ and $\beta/(1 - \alpha)$ are very nearly equal to α and β, respectively. Thus, also in case of grouping, the realized values α' and β' cannot exceed the intended values α and β, respectively, except by an exceedingly small quantity which can be neglected for all practical purposes. This means that, for all practical purposes, grouping will not decrease the protection against wrong decisions provided by the test. The only possible effect of practical significance that may be caused by grouping is that it may make α' or β' substantially smaller than the intended values α and β. This feature of grouping compensates, to some extent, for the increase in the number of observations.

It may be of interest to remark that, if the number v of units in a group is equal to the reciprocal of the common slope s of the acceptance and rejection lines and if the intercepts of these lines are integers, the OC curve is not affected at all by grouping.[13] This can be seen as follows: Because $s = 1/d$, we have $a_{m+d} = a_m + 1$ and $r_{m+d} = r_m + 1$. Furthermore, since the intercepts of the acceptance and rejection lines are assumed to be integers, a_m and r_m have integral values for any m which is an integral multiple of v. If item-by-item inspection leads to acceptance of the lot at the nth item, then n must be an integral multiple of v, and therefore inspection in groups of v will also lead to acceptance. If item-by-item inspection leads to rejection of the lot at the nth item, then we have $d_n \geq r_n$. Let n' be the smallest integral multiple of v greater than or equal to n. Then $d_n = r_{n'}$, since d_n is an integer, $d_n - r_n \leq 1$, and $r_{n'} - r_n \leq 1$. Hence $d_{n'} \geq r_{n'}$ and inspection in groups will also terminate with rejection of the lot. Thus, inspection in groups leads to exactly the same decision as item-by-item inspection and consequently grouping does not affect the OC curve.

5.6.2 Upper and Lower Limits for the Effect of Grouping on the OC and ASN Curves

Upper and lower limits for the effect of grouping on the OC and ASN curves can be obtained by considering the following three auxiliary sequential sampling plans. Let h_0 be the intercept of the acceptance line, h_1 the intercept of the rejection line, and s the common slope in the given sampling plan. The first auxiliary plan is obtained by changing h_0 to $h_0^* = h_0 - vs$ and leaving h_1 and s unchanged. The second auxiliary plan is obtained by changing h_1 to $h_1^* = h_1 + vs$, leaving h_0 and s unchanged. Finally, the third auxiliary plan corresponds to the intercepts h_0^*, h_1^*, and slope s. Let $L_i(p)$ denote the OC function and $E_{pi}(n)$ the ASN function of the auxiliary plan i, when item-by-item inspection is used ($i = 1, 2, 3$). Furthermore, let $L(p)$ denote the OC function and $E_p(n)$ the ASN function of the given plan when item-by-item inspection is used. When inspection is made in groups the OC and ASN functions are affected,[14] and we shall denote them by $\bar{L}(p)$ and $\bar{E}_p(n)$ respectively.

[13] See also SRG 255, p. 2.30.

[14] Except, in the case of the OC function, when the number of units in the group is the reciprocal of the slope, as stated in Section 5.6.1.

It can easily be seen that whenever the first auxiliary plan (using item-by-item inspection) leads to the acceptance of the lot, the original plan (taking observations in groups) also leads to acceptance. The converse is, however, not necessarily true. That is, it may happen that the auxiliary plan leads to rejection of the lot, whereas the original plan leads to acceptance. Thus, we have

$$(5:32) \qquad\qquad L_1(p) \leq \bar{L}(p)$$

Similarly, one can verify that whenever the second auxiliary plan (using item-by-item inspection) leads to rejection of the lot, the original plan (using grouping) also leads to rejection. Hence

$$(5:33) \qquad\qquad 1 - L_2(p) \leq 1 - \bar{L}(p)$$

This can be written as

$$(5:34) \qquad\qquad \bar{L}(p) \leq L_2(p)$$

From (5:32) and (5:34) we obtain

$$(5:35) \qquad\qquad L_1(p) \leq \bar{L}(p) \leq L_2(p)$$

To derive an upper limit for $\bar{E}_p(n)$ we shall make use of the third auxiliary plan. If this plan (using item-by-item inspection) terminates at the inspection of the nth unit, the original plan (using grouping) must terminate at the latest with the inspection of the group in which the nth item is included.[15] Hence, the number n' of units inspected when the original plan is used cannot exceed $n + v$. From this it follows that

$$(5:36) \qquad\qquad \bar{E}_p(n) \leq E_{p3}(n) + v$$

Since $E_p(n) \leq \bar{E}_p(n)$, we obtain the limits

$$(5:37) \qquad\qquad E_p(n) \leq \bar{E}_p(n) \leq E_{p3}(n) + v$$

Limits for $\bar{L}(p)$ and $\bar{E}_p(n)$ could also be derived by using the method described in Sections A.2.3 and A.3.1 of the Appendix. The limits given in (5:35) and (5:37) will be rather close when p_1/p_0 and $(1 - p_1)/(1 - p_0)$ are near 1 and vs does not exceed 1.

5.7   Truncation of the Test Procedure

The sequential sampling plan does not provide any definite upper bound for the number n of units to be inspected. Any large value of n is possible, but the probability is small that n will exceed twice or

---

[15] It is possible, of course, that inspection terminates with an earlier group.

three times its expected value. It is sometimes desirable to set a definite upper bound n_0 for n, excluding even a small probability that n may exceed n_0. This can be done by truncating the sequential process at $n = n_0$. That is to say, we terminate the process at $n = n_0$ even if the regular sequential rule does not lead to a final decision for $n \leq n_0$. The following seems to be a reasonable rule for deciding acceptance or rejection of the lot at $n = n_0$ if no decision is reached for $n \leq n_0$ with the regular sequential procedure: If $d_{n_0} \geq (a_{n_0} + r_{n_0})/2$ we reject the lot, and if $d_{n_0} < (a_{n_0} + r_{n_0})/2$ we accept the lot.

Truncation and its effect on the OC curve are discussed in Section 3.8. If n_0 is put as high as three times the expected value of n, the effect of truncation on the OC curve is negligibly small, since the probability is nearly 1 that the regular sequential procedure will terminate for $n < n_0$.

Chapter 6. **TESTING THE DIFFERENCE BETWEEN THE MEANS OF TWO BINOMIAL DISTRIBUTIONS (DOUBLE DICHOTOMIES)**

6.1 Formulation of the Problem

Suppose that we want to compare the effectiveness of two production processes where the effectiveness of a production process is measured in terms of the proportion of effective units in the sequence produced. We shall say that a unit is effective if it has a certain desirable property, for example, if it withstands a certain strain. Let p_1 be the proportion of effectives if process 1 is used, and p_2 the proportion of effectives if process 2 is used. In other words, p_1 is the probability that a unit produced will be effective if process 1 is used, and p_2 is the probability that a unit produced will be effective if process 2 is used. Suppose that the manufacturer does not know the values of p_1 and p_2, and that process 1 is in operation. If $p_1 \geqq p_2$, the manufacturer wants to retain process 1. However, if $p_1 < p_2$, especially if p_1 is substantially smaller than p_2, the manufacturer would like to replace process 1 by process 2. Thus, we are interested in testing the hypothesis that $p_1 \geqq p_2$ against the alternative that $p_1 < p_2$.

A more general formulation of the problem can be stated as follows. Consider two binomial distributions. Let p_1 be the probability of a success in a single trial according to the first binomial distribution, and let p_2 be the probability of a success in a single trial according to the second binomial distribution. We shall use the symbol 1 for success and the symbol 0 for failure. Suppose that the probabilities p_1 and p_2 are unknown. We consider the problem of testing the hypothesis that $p_1 \geqq p_2$ on the basis of a sample consisting of N_1 observations from the first binomial distribution and N_2 observations from the second binomial distribution. Since in many experiments the case $N_1 = N_2$ is mainly of interest, and since this case (as we shall see later) makes an exact and simplified mathematical treatment of the problem possible, in what follows we shall assume that $N_1 = N_2 = N$ (say). Thus, on the basis of the outcome of the two series of N independent trials we have to decide whether the hypothesis $p_1 \geqq p_2$ should be accepted or rejected.

6.2 The Classical Method

The classical solution of the problem for large N is given as follows. Let S_1 be the number of successes in the first set of N trials (drawn from the first binomial distribution), and let S_2 be the number of successes in the second set of N trials (drawn from the second binomial population). Denote $(S_1 + S_2)/2N$ by $\bar{p}$ and $1 - \bar{p}$ by $\bar{q}$. Then for large N the expression

$$(6:1) \qquad \frac{S_2 - S_1}{\sqrt{2N\bar{p}\bar{q}}}.$$

is normally distributed with zero mean and unit variance if $p_1 = p_2$. Suppose that the level of significance we wish to choose is α. Let λ_α be the value for which the probability that a normal variate with zero mean and unit variance will exceed λ_α is equal to α. (For example, if $\alpha = .05$, $\lambda_\alpha = 1.64$.) Thus, if $p_1 = p_2$, the probability that the expression (6:1) will exceed λ_α is equal to α. If $p_1 > p_2$, the probability that the expression (6:1) will exceed λ_α is less than α. According to the classical method, the hypothesis that $p_1 \geqq p_2$ is rejected if the observed value of (6:1) exceeds λ_α. This method involves an approximation, since the distribution of (6:1) is not exactly normal (for small N it is far from normal). For small N an exact method has been proposed by R. A. Fisher which, however, involves cumbersome calculations. In Section 6.3 we shall suggest another (non-sequential) method which is exact and is fairly simple to apply as far as computations are concerned. The latter method has the further advantage of being suitable for sequential analysis, to which existing methods are not readily adaptable.

6.3 An Exact Non-Sequential Method

Let $a_1, \cdots, a_N$ be the results in the first set of N trials, and $b_1, \cdots, b_N$ the results in the second set of N trials. These results are arranged in the order observed. Consider the sequence of N pairs:

$$(6:2) \qquad (a_1, b_1), \cdots, (a_N, b_N)$$

Let t_1 be the number of pairs $(1, 0)$ and t_2 the number of pairs $(0, 1)$ in this sequence. We consider only the pairs $(0, 1)$ and $(1, 0)$ and base the test on them.

Let a be the outcome of an observation from the first population, and b the outcome of an observation from the second population. The probability that $(a, b) = (1, 0)$ is equal to $p_1(1 - p_2)$, and the probability that $(a, b) = (0, 1)$ is equal to $(1 - p_1)p_2$. Hence, knowing that (a, b) is equal to one of the pairs $(0, 1)$ and $(1, 0)$, the (conditional) probability that it is equal to $(0, 1)$ is given by

$$(6:3) \qquad p = \frac{(1 - p_1)p_2}{p_1(1 - p_2) + p_2(1 - p_1)}$$

and the (conditional) probability that it is equal to $(1, 0)$ is given by

$$(6:4) \qquad 1 - p = \frac{p_1(1 - p_2)}{p_1(1 - p_2) + (1 - p_1)p_2}$$

Hence, when only the pairs $(1, 0)$ and $(0, 1)$ are considered, the variate t_2 is distributed like the number of successes in a sequence of $t = t_1 + t_2$ independent trials, the probability of a success in a single trial being equal to p. One can easily verify that $p = \frac{1}{2}$ if $p_1 = p_2$, $p < \frac{1}{2}$ if $p_1 > p_2$, and $p > \frac{1}{2}$ if $p_1 < p_2$. Thus, the hypothesis to be tested, i.e., the hypothesis that $p_1 \geq p_2$, is equivalent to the hypothesis that $p \leq \frac{1}{2}$. Thus, we can test the hypothesis that $p_1 \geq p_2$ by testing the hypothesis that $p \leq \frac{1}{2}$ on the basis of the observed value of t_2. Since the distribution of t_2 is the same as the distribution of the number of successes in $t = t_1 + t_2$ independent trials (t is treated as a constant and the probability of a success in a single trial is equal to p), the test procedure can be carried out in the usual manner. If we want a level of significance α, a critical value T is chosen so that for $p = \frac{1}{2}$ the probability that $t_2 \geq T$ is equal to α. The hypothesis that $p \leq \frac{1}{2}$ is rejected if and only if the observed t_2 is greater than or equal to the critical value T. The value of T can be obtained from a table of the binomial distribution. If t is large, t_2 is nearly normally distributed, and the critical value T can be obtained from a table of the normal distribution.

This procedure thus provides a simple test of the hypothesis that $p_1 \geq p_2$. The question arises whether the efficiency of this method is as high as that of the classical method. It would seem that the method suggested here cannot be a most efficient procedure, since the values of t_1 and t_2 depend on the order of the elements in the sequences $(a_1, \cdots, a_N)$ and $(b_1, \cdots, b_N)$, and there is no particular reason to arrange them in the order observed. However, it has been shown [1] that the loss in efficiency as compared with the classical method is negligible if the number N of trials is large.[2]

[1] See the author's report, *Sequential Analysis of Statistical Data: Theory*, submitted to the Applied Mathematics Panel, National Defense Research Committee, Sept., 1943.

[2] The author believes that the loss in efficiency is slight even when N is small, although no exact investigation of this case has been made.

It should be pointed out that the procedure for testing the hypothesis that $p_1 \geqq p_2$ can be used also for testing the hypothesis that $p_1 = p_2$ if the alternative hypotheses are restricted to $p_2 > p_1$.

In addition to simplicity and exactness, the present method seems superior to the classical one in the following respect. Suppose that (contrary to the original assumption) the probability of a success varies from trial to trial. Let $p_1^{(i)}$ denote the probability of success in the ith trial of the first set, and let $p_2^{(i)}$ denote the probability of success in the ith trial in the second set $(i = 1, \cdots, N)$. Assume that the probabilities $p_1^{(i)}$ and $p_2^{(i)}$ are entirely unknown and we wish to test the hypothesis that $p_1^{(1)} - p_2^{(1)} = \cdots = p_1^{(N)} - p_2^{(N)} = 0$. In this case the classical method is not applicable, but the present method provides a correct procedure. Such a situation may arise, for instance, if we want to test the hypothesis that the probability of a success (hitting the target) is the same for two different guns. In the course of the experiments the probability of a hit may change because of external conditions such as wind or disposition of the gunner. However, these external conditions are likely to affect both guns equally if the trials are made alternately (or approximately alternately), so that if the two guns are equally good we have $p_1^{(i)} = p_2^{(i)}$ $(i = 1, \cdots N)$.

6.4   Sequential Test of the Hypothesis That $p_1 \geqq p_2$

6.4.1   Risks That We Are Willing to Tolerate of Making Wrong Decisions

In order to devise a proper sequential test for testing the hypothesis that $p_1 \geqq p_2$, we have to state first what risks of making wrong decisions we are willing to tolerate. The efficiency of production process 1 may be measured by the ratio of effectives to ineffectives produced, i.e., by $k_1 = p_1/(1 - p_1)$. Production process 1 may be regarded the more efficient the larger the value of k_1. Similarly, the efficiency of production process 2 may be measured by $k_2 = p_2/(1 - p_2)$. The relative superiority of production process 2 over process 1 can then reasonably be measured by the ratio of k_2 to k_1, i.e., by

$$(6:5) \qquad u = \frac{k_2}{k_1} = \frac{p_2(1 - p_1)}{p_1(1 - p_2)}$$

If $u = 1$, the two processes are equally good. If $u > 1$, process 2 is superior to process 1, and if $u < 1$, process 1 is superior to process 2. Thus, the manufacturer will, in general, be able to select two values of u, u_0 and u_1, say $(u_0 < u_1)$, such that the rejection of process 1 in favor of process 2 is considered an error of practical importance whenever the true value of $u \leqq u_0$, and the maintenance of process 1 is considered an error of practical importance whenever $u \geqq u_1$. If u lies between u_0 and u_1, the manufacturer does not care particularly which decision is made.

Clearly, we will always have $u_0 < u_1$. If the transition from production process 1 to process 2 involves some cost or other inconveniences, it seems reasonable to put $u_0 = 1$ (or u_0 may even be slightly greater than 1). This choice of u_0 really means that we consider the rejection of process 1 a serious error whenever this process is not inferior to process 2. On the other hand, if the transition from process 1 to process 2 does not involve any inconveniences, the rejection of process 1 in favor of 2 cannot be a serious error when the two processes are equally efficient, i.e., when $u = 1$. Thus, in such a case it seems reasonable to choose u_0 somewhat below 1.

After the quantities u_0 and u_1 have been chosen the risks that we are willing to tolerate may reasonably be expressed in the following form: The probability of rejecting process 1 should not exceed a preassigned value α whenever $u \leqq u_0$, and the probability of maintaining process 1 should not exceed a preassigned value β whenever $u \geqq u_1$. Thus, the risks that we are willing to tolerate are characterized by the four quantities u_0, u_1, α, and β.

6.4.2   The Sequential Probability Ratio Test Corresponding to the Quantities u_0, u_1, α, and β

After the four quantities u_0, u_1, α, and β have been chosen, a proper sequential test can be carried out as follows. The (conditional) probability that we obtain a pair $(0, 1)$, as given in $(6\!:\!3)$, can be expressed as a function of u. In fact

$$(6\!:\!6) \quad p = \frac{(1 - p_1)p_2}{p_1(1 - p_2) + p_2(1 - p_1)} = \frac{\dfrac{(1 - p_1)p_2}{p_1(1 - p_2)}}{1 + \dfrac{p_2(1 - p_1)}{p_1(1 - p_2)}} = \frac{u}{1 + u}$$

Let H_0 denote the hypothesis that $p = u_0/(1 + u_0)$, and H_1 the hypothesis that $p = u_1/(1 + u_1)$. A proper sequential test satisfying our requirements concerning tolerated risks is the sequential probability ratio test of H_0 against H_1. The acceptance and rejection numbers for this sequential test can be obtained from $(5\!:\!11)$ and $(5\!:\!12)$ by substituting $u_0/(1 + u_0)$ for p_0, $u_1/(1 + u_1)$ for p_1, and $t = t_1 + t_2$ for m.

Thus, for each value of t the acceptance number is given by

$$(6\!:\!7) \qquad a_t = \frac{\log \dfrac{\beta}{1-\alpha}}{\log u_1 - \log u_0} + t\, \frac{\log \dfrac{1+u_1}{1+u_0}}{\log u_1 - \log u_0}$$

and the rejection number is given by

$$(6\!:\!8) \qquad r_t = \frac{\log \dfrac{1-\beta}{\alpha}}{\log u_1 - \log u_0} + t\, \frac{\log \dfrac{1+u_1}{1+u_0}}{\log u_1 - \log u_0}$$

These acceptance numbers a_t and rejection numbers r_t $(t = 1, 2, \cdots)$ are best tabulated before experimentation starts. The sequential test is then carried out as follows. The observations are taken in pairs where each pair consists of an observation from the first process and an observation from the second process. We continue taking pairs as long as $a_t < t_2 < r_t$. The first time that t_2 does not lie between the acceptance and rejection numbers, experimentation is terminated. Process 1 is maintained if at this final stage $t_2 \leqq a_t$, and process 1 is rejected in favor of 2 if $t_2 \geqq r_t$.

As an illustration, the following example is given. Let $u_0 = 1.3$, $u_1 = 3$, $\alpha = .03$, and $\beta = .10$. The observed pairs $(0, 1)$ and $(1, 0)$ in an experiment, and the rejection and acceptance numbers, are given in Table 6. In this example, the sampling process is terminated at $t = 18$ with the retention of process 1.

The test procedure can also be carried out graphically as shown in Fig. 14. The total number t of pairs $(0, 1)$ and $(1, 0)$ is measured along

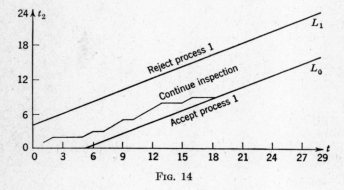

FIG. 14

DOUBLE DICHOTOMIES

the horizontal axis. The points (t, a_t) will lie on a straight line L_0, since a_t is a linear function of t. The points (t, r_t) will lie on a parallel line L_1. We draw the lines L_0 and L_1 and plot the points (t, t_2) as

TABLE 6

t Number of Pairs (0, 1), (1, 0) Observed	Pairs (0, 1), (1, 0) Observed	a_t Acceptance Number	t_2 Number of Pairs (0, 1) Observed	r_t Rejection Number
1	(0, 1)	..	1	..
2	(0, 1)	..	2	..
3	(1, 0)	..	2	..
4	(1, 0)	..	2	..
5	(1, 0)	0	2	..
6	(0, 1)	1	3	..
7	(1, 0)	1	3	..
8	(0, 1)	2	4	..
9	(0, 1)	3	5	..
10	(1, 0)	3	5	..
11	(0, 1)	4	6	..
12	(0, 1)	5	7	..
13	(0, 1)	5	8	13
14	(1, 0)	6	8	14
15	(1, 0)	7	8	14
16	(0, 1)	7	9	15
17	(1, 0)	8	9	16
18	(1, 0)	9	9	16
19		9		17
20		10		18
21		11		18
22		11		19
23		12		20
24		13		20
25		13		21
26		14		22
27		15		22
28		15		23
29		16		24

experimentation goes on. The first time that the point (t, t_2) is not within the lines L_0 and L_1 experimentation is terminated. Process 1 is maintained if at the final stage (t, t_2) lies on L_0 or below, and process 1 is rejected if (t, t_2) lies on L_1 or above.

The intercept of line L_0 is given by

$$(6:9) \qquad h_0 = \frac{\log \dfrac{\beta}{1 - \alpha}}{\log u_1 - \log u_0}$$

and the intercept of L_1 is given by

$$(6:10) \qquad h_1 = \frac{\log \dfrac{1 - \beta}{\alpha}}{\log u_1 - \log u_0}$$

The common slope of the two lines is equal to

$$(6:11) \qquad s = \frac{\log \dfrac{1 + u_1}{1 + u_0}}{\log u_1 - \log u_0}$$

6.4.3 The Operating Characteristic Curve of the Test

For any value u of the ratio k_2/k_1, we shall denote by $L(u)$ the probability of maintaining process 1. Clearly, $L(u)$ is a function of u. This function $L(u)$ is called the operating characteristic function of the test. It can be obtained from equations (5:19) and (5:20) by substituting $u_0/(1 + u_0)$ for p_0 and $u_1/(1 + u_1)$ for p_1. These equations are: [3]

$$(6:12) \qquad L(u) = \frac{\left(\dfrac{1 - \beta}{\alpha}\right)^h - 1}{\left(\dfrac{1 - \beta}{\alpha}\right)^h - \left(\dfrac{\beta}{1 - \alpha}\right)^h}$$

and

$$(6:13) \qquad \frac{u}{1 + u} = \frac{1 - \left(\dfrac{1 + u_0}{1 + u_1}\right)^h}{\left(\dfrac{u_1(1 + u_0)}{u_0(1 + u_1)}\right)^h - \left(\dfrac{1 + u_0}{1 + u_1}\right)^h}$$

Equation (6:13) can be written as

$$(6:14) \qquad u = \frac{1 - \left(\dfrac{1 + u_0}{1 + u_1}\right)^h}{\left(\dfrac{u_1(1 + u_0)}{u_0(1 + u_1)}\right)^h - 1}$$

[3] In the formulas given in SRG 255, p. 3.38, the quantities u and $L(u)$ are expressed in terms of a "dummy" variable x which is functionally related to h.

For any given value h we compute u and $L(u)$ from the equations (6:12) and (6:14). The point $[u, L(u)]$ obtained in this way will be a point of the OC curve. By calculating the points $[u, L(u)]$ for a sufficiently large number of values of h, the OC curve can be drawn.

We shall compute $[u, L(u)]$ for $h = -\infty, -1, 0, 1, +\infty$. Since $\dfrac{1 + u_0}{1 + u_1} < 1$ and $\dfrac{u_1(1 + u_0)}{u_0(1 + u_1)} > 1$, we obtain from (6:12) and (6:14)

(6:15) $u = \infty$ and $L(u) = 0$ when $h = -\infty$

(6:16) $u = 0$ and $L(u) = 1$ when $h = +\infty$

Furthermore we obtain

(6:17) $u = u_1$ and $L(u) = \beta$ when $h = -1$

and

(6:18) $u = u_0$ and $L(u) = 1 - \alpha$ when $h = +1$

For $h = 0$, the expressions u and $L(u)$ have the form $0/0$. The limiting values of u and $L(u)$ when $h \to 0$ can be obtained by differentiating numerator and denominator at $h = 0$. Then we have

$$(6:19) \quad u = \frac{\log \dfrac{1 + u_1}{1 + u_0}}{\log \dfrac{u_1(1 + u_0)}{u_0(1 + u_1)}} \quad \text{and} \quad L(u) = \frac{\log \dfrac{1 - \beta}{\alpha}}{\log \dfrac{1 - \beta}{\alpha} - \log \dfrac{\beta}{1 - \alpha}}$$

when $h = 0$.

These five points on the OC curve already determine roughly the shape of the curve. It can be seen that u is a decreasing function of h and $L(u)$ is an increasing function of h. Hence $L(u)$ is a decreasing function of u. As u varies from 0 to u_0, $L(u)$ decreases from 1 to $1 - \alpha$. In the interval from u_0 to u_1, $L(u)$ decreases from $1 - \alpha$ to β, and as u varies from u_1 to $+\infty$, the OC function $L(u)$ decreases from β to 0.

6.4.4   The Average Amount of Inspection Required by the Test

For any value u of the ratio k_2/k_1, let $E_u(t)$ denote the expected value of the total number of pairs $(0, 1)$ and $(1, 0)$ required by the test. The value of $E_u(t)$ can be obtained from (5:23) by substituting $E_u(t)$ for $E_p(n)$, $L(u)$ for $L(p)$, $u_0/(1 + u_0)$ for p_0, $u_1/(1 + u_1)$ for p_1, and $u/(1 + u)$ for p. Thus [4]

---

[4] The right-hand member of (6:20) can be expressed as a function of $L(u)$, the intercepts and the slope of the decision lines. See SRG 255, p. 3.41.

$$(6:20) \quad E_u(t) = \frac{L(u) \log \dfrac{\beta}{1-\alpha} + (1 - L(u)) \log \dfrac{1-\beta}{\alpha}}{\dfrac{u}{1+u} \log \dfrac{u_1(1+u_0)}{u_0(1+u_1)} + \dfrac{1}{1+u} \log \dfrac{1+u_0}{1+u_1}}$$

To compute the expected value of the total number of pairs (including also the pairs $(0, 0)$ and $(1, 1)$), we merely have to divide the right-hand side of equation $(6:20)$ by $p_1(1 - p_2) + p_2(1 - p_1)$.

Since $L(0) = 1$ and $L(\infty) = 0$, we obtain from $(6:20)$

$$(6:21) \quad E_u(t) = \frac{\log \dfrac{\beta}{1-\alpha}}{\log \dfrac{1+u_0}{1+u_1}} \quad \text{when } u = 0$$

and

$$(6:22) \quad E_u(t) = \frac{\log \dfrac{1-\beta}{\alpha}}{\log \dfrac{u_1(1+u_0)}{u_0(1+u_1)}} \quad \text{when } u = \infty$$

Since $L(u_0) = 1 - \alpha$ and $L(u_1) = \beta$, it follows from $(6:20)$ that

$$(6:23) \quad E_{u_0}(t) = \frac{(1-\alpha) \log \dfrac{\beta}{1-\alpha} + \alpha \log \dfrac{1-\beta}{\alpha}}{\dfrac{u_0}{1+u_0} \log \dfrac{u_1(1+u_0)}{u_0(1+u_1)} + \dfrac{1}{1+u_0} \log \dfrac{1+u_0}{1+u_1}}$$

and

$$(6:24) \quad E_{u_1}(t) = \frac{\beta \log \dfrac{\beta}{1-\alpha} + (1-\beta) \log \dfrac{1-\beta}{\alpha}}{\dfrac{u_1}{1+u_1} \log \dfrac{u_1(1+u_0)}{u_0(1+u_1)} + \dfrac{1}{1+u_1} \log \dfrac{1+u_0}{1+u_1}}$$

In Section 5.5 we have computed the expected value of n when p is equal to the slope of the acceptance and rejection lines. This corresponds to the case when $u/(1 + u) = s$, i.e., $u = s/(1 - s)$, where the slope s is given in $(6:11)$. The value of $E_u(t)$ for $u = s/(1 - s)$ can be obtained from the right-hand member of $(5:30)$, replacing p_1 by $u_1/(1 + u_1)$ and p_0 by $u_0/(1 + u_0)$. Thus

$$(6:25) \quad E_{\frac{s}{1-s}}(t) = \frac{-\left(\log \dfrac{\beta}{1-\alpha}\right)\left(\log \dfrac{1-\beta}{\alpha}\right)}{\log \dfrac{u_1(1+u_0)}{u_0(1+u_1)} \log \dfrac{1+u_1}{1+u_0}}$$

The determination of the five values of $E_u(t)$, as given in (6:21) through (6:25), may frequently suffice in practice, since these five points generally give a fairly good idea of the shape of the whole curve.

6.4.5   Observations Taken in Groups

In applications it may happen that, at each stage in the sequential process, instead of drawing a single observation we draw a group of v observations from each of the binomial distributions. Hence, instead of a single pair, we have two groups of v observations. The effect of grouping on the OC and ASN curves has been discussed in Section 5.6 and the results obtained there can be applied to the case under consideration here. If the order of observations in each group of v is recorded, we can establish the number of pairs $(0, 1)$ and the number of pairs $(1, 0)$ for each pair of groups of v observations. In such a case the test can be carried out as described in Section 6.4.2, since after each pair of groups of v observations we can compute t and t_2. However, if the order of observations in such groups is not recorded, the difficulty arises that we are not able to determine the values of t and t_2 needed for the test procedure.

It has been shown [5] that in such a case we may replace t and t_2 by certain estimates of t and t_2 without affecting seriously the probability of making an incorrect decision. The estimates of t_1 and t_2 (and thereby also an estimate of $t = t_1 + t_2$) are obtained as follows. Let v_1 be the number of successes in the group of v observations drawn from the first binomial distribution, and let v_2 be the number of successes in the group of v observations drawn from the second binomial distribution. Then for this pair of groups of v observations we estimate the number of pairs $(1, 0)$ to be $v_1 - (v_1 v_2/v)$ and the number of pairs $(0, 1)$ to be $v_2 - (v_1 v_2/v)$. Thus, an estimate of t_1 is obtained by summing $v_1 - (v_1 v_2/v)$ over all pairs of groups observed, and that of t_2 is obtained by summing $v_2 - (v_1 v_2/v)$ over all pairs of groups observed.

For the effect of grouping on the OC and ASN curves, the results of Section 5.6 can be applied, since the test procedure discussed here reduces to that considered in Section 5.6 when $p = u/(1 + u)$, $m = t_1 + t_2 = t$, and $d_m = t_2$.

[5] See the author's report, *Sequential Analysis of Statistical Data: Theory*, submitted to the Applied Mathematics Panel, National Defense Research Committee, Sept., 1943.

Chapter 7. TESTING THAT THE MEAN OF A NORMAL DISTRIBUTION WITH KNOWN STANDARD DEVIATION FALLS SHORT OF A GIVEN VALUE

7.1 Formulation of the Problem

Let x be a random variable which is normally distributed with unknown mean θ and known standard deviation σ. In this section we shall deal with the problem of testing the hypothesis that θ is less than or equal to some specified value θ'.

Such a problem arises frequently, for example, in quality control and acceptance inspection. Suppose that a lot consisting of a large number of units of a manufactured product is submitted for acceptance inspection. The number of units in the lot is assumed to be sufficiently large so that the lot may be treated as containing infinitely many units. Suppose that the result of an observation is a measurement x of some quality characteristic of the unit, such as the weight, or hardness, or tensile strength. The value of x will, in general, vary from unit to unit. It is assumed that x is normally distributed with a known standard deviation σ but unknown mean θ. Suppose, furthermore, that the product is considered the more desirable the smaller the value of θ. Then it will, in general, be possible to designate a particular value θ' such that we prefer to accept the lot if $\theta < \theta'$ and we prefer to reject the lot if $\theta > \theta'$. Thus, in such a situation, we are interested in devising a sampling plan to test the hypothesis that $\theta < \theta'$.

Since quality control and acceptance inspection is an important field of application for such test procedures, we shall continue the discussion using the terminology of acceptance inspection. This, of course, should not be interpreted as a restriction on the general validity and applicability of the test procedure.

7.2 Tolerated Risks of Making Wrong Decision

If $\theta = \theta'$, we are indifferent whether the lot is accepted or rejected. The preference for acceptance increases with decreasing value of θ in the domain $\theta < \theta'$, and the preference for rejection increases with increasing value of θ in the domain $\theta > \theta'$. Thus, it will be possible, in general, to find two values θ_0 and θ_1 ($\theta_0 < \theta'$ and $\theta_1 > \theta'$) such that rejection of the lot is considered an error of practical consequence if $\theta \leq \theta_0$, and acceptance of the lot is considered an error of practical consequence if $\theta \geq \theta_1$; for values θ between θ_0 and θ_1 we do not care particularly which decision is taken. Using the terminology introduced in Section 2.3.1, we may say that the zone of preference for acceptance consists of all values θ for which $\theta \leq \theta_0$, the zone of preference for rejection is the set of all values θ for which $\theta \geq \theta_1$, and the zone of indifference consists of all values θ between θ_0 and θ_1.

After the two values θ_0 and θ_1 have been chosen the risks that we are willing to tolerate may reasonably be expressed as follows.[1] The probability of rejecting the lot should not exceed a small preassigned value α whenever $\theta \leq \theta_0$, and the probability of accepting the lot should not exceed a small preassigned value β whenever $\theta \geq \theta_1$. Thus, the risks that we are willing to tolerate are characterized by the four numbers θ_0, θ_1, α, and β.

7.3   The Sequential Probability Ratio Test Corresponding to the Quantities θ_0, θ_1, α, and β

The requirements regarding the tolerated risks are satisfied by the sequential probability ratio test of strength (α, β) for testing the hypothesis that $\theta = \theta_0$ against the alternative that $\theta = \theta_1$. This sequential test is given as follows. Let $x_1, x_2, \cdots$, etc., be the successive observations on x. The probability density of the sample $(x_1, \cdots, x_m)$ is given by

$$(7:1) \qquad p_{0m} = \frac{1}{(2\pi)^{\frac{m}{2}}\sigma^m} e^{-\frac{1}{2\sigma^2}\sum_{\alpha=1}^{m}(x_\alpha - \theta_0)^2}$$

if $\theta = \theta_0$, and by

$$(7:2) \qquad p_{1m} = \frac{1}{(2\pi)^{\frac{m}{2}}\sigma^m} e^{-\frac{1}{2\sigma^2}\sum_{\alpha=1}^{m}(x_\alpha - \theta_1)^2}$$

if $\theta = \theta_1$. The probability ratio p_{1m}/p_{0m} is computed at each stage of the inspection. Additional observations are taken as long as

$$(7:3) \qquad B < \frac{p_{1m}}{p_{0m}} = \frac{e^{-\frac{1}{2\sigma^2}\Sigma(x_\alpha - \theta_1)^2}}{e^{-\frac{1}{2\sigma^2}\Sigma(x_\alpha - \theta_0)^2}} < A$$

[1] See, for instance, Section 2.3.2.

Inspection is terminated with the acceptance of the lot if

$$(7:4) \qquad \frac{e^{-\frac{1}{2\sigma^2}\Sigma(x_\alpha-\theta_1)^2}}{e^{-\frac{1}{2\sigma^2}\Sigma(x_\alpha-\theta_0)^2}} \leqq B$$

Inspection is terminated with the rejection of the lot if

$$(7:5) \qquad \frac{e^{-\frac{1}{2\sigma^2}\Sigma(x_\alpha-\theta_1)^2}}{e^{-\frac{1}{2\sigma^2}\Sigma(x_\alpha-\theta_0)^2}} \geqq A$$

According to Section 3.3 approximate values of A and B are given by $(1 - \beta)/\alpha$ and $\beta/(1 - \alpha)$, respectively.

By taking logarithms and simplifying, the inequalities (7:3), (7:4), and (7:5) can be written as

$$(7:6) \quad \log \frac{\beta}{1-\alpha} < \frac{\theta_1 - \theta_0}{\sigma^2} \sum_{\alpha=1}^{m} x_\alpha + \frac{m}{2\sigma^2}(\theta_0{}^2 - \theta_1{}^2) < \log \frac{1-\beta}{\alpha}$$

$$(7:7) \qquad \frac{\theta_1 - \theta_0}{\sigma^2} \sum_{\alpha=1}^{m} x_\alpha + \frac{m}{2\sigma^2}(\theta_0{}^2 - \theta_1{}^2) \leqq \log \frac{\beta}{1-\alpha}$$

and

$$(7:8) \qquad \frac{\theta_1 - \theta_0}{\sigma^2} \sum_{\alpha=1}^{m} x_\alpha + \frac{m}{2\sigma^2}(\theta_0{}^2 - \theta_1{}^2) \geqq \log \frac{1-\beta}{\alpha}$$

respectively.

Further simplification in carrying out the test procedure can be achieved by adding $(-m/2\sigma^2)(\theta_0{}^2 - \theta_1{}^2)$ to both sides of the inequalities (7:6), (7:7), and (7:8) and then dividing these inequalities by $(\theta_1 - \theta_0)/\sigma^2$. These operations transform the inequalities (7:6), (7:7), and (7:8) into

$$(7:9) \quad \frac{\sigma^2}{\theta_1 - \theta_0} \log \frac{\beta}{1-\alpha} + m \frac{\theta_0 + \theta_1}{2} <$$

$$\sum_{\alpha=1}^{m} x_\alpha < \frac{\sigma^2}{\theta_1 - \theta_0} \log \frac{1-\beta}{\alpha} + m \frac{\theta_0 + \theta_1}{2}$$

$$(7:10) \qquad \Sigma x_\alpha \leqq \frac{\sigma^2}{\theta_1 - \theta_0} \log \frac{\beta}{1-\alpha} + m \frac{\theta_0 + \theta_1}{2}$$

and

$$(7:11) \qquad \Sigma x_\alpha \geqq \frac{\sigma^2}{\theta_1 - \theta_0} \log \frac{1-\beta}{\alpha} + m \frac{\theta_0 + \theta_1}{2}$$

respectively.

By using the inequalities (7:9), (7:10), and (7:11) the inspection plan may be carried out as follows. For each m compute the acceptance number

$$(7:12) \qquad a_m = \frac{\sigma^2}{\theta_1 - \theta_0} \log \frac{\beta}{1 - \alpha} + m \frac{\theta_0 + \theta_1}{2}$$

and the rejection number

$$(7:13) \qquad r_m = \frac{\sigma^2}{\theta_1 - \theta_0} \log \frac{1 - \beta}{\alpha} + m \frac{\theta_0 + \theta_1}{2}$$

These acceptance and rejection numbers are best computed before inspection starts. Inspection is continued as long as $a_m < \Sigma x_\alpha < r_m$. At the first time when Σx_α does not lie between a_m and r_m, inspection is terminated. The lot is accepted if $\Sigma x_\alpha \leqq a_m$, and the lot is rejected if $\Sigma x_\alpha \geqq r_m$.

As an illustration, consider the following example. Let $\theta_0 = 135$, $\theta_1 = 150$, $\alpha = .01$, and $\beta = .03$. Furthermore, let $\sigma = 25$. The observations and the acceptance and rejection numbers are tabulated in Table 7, which shows that the sampling inspection is terminated at $m = 20$ with the acceptance of the lot.

The test procedure can also be carried out graphically as shown in Fig. 15. The number m of observations is measured along the hori-

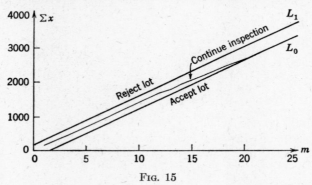

Fig. 15

zontal axis. The points (m, a_m) will lie on a straight line L_0 and the points (m, r_m) will lie on a parallel line L_1. We draw the parallel lines L_0 and L_1 before inspection starts. The points $(m, \sum_{\alpha=1}^{m} x)$ are plotted as inspection goes on. Inspection is continued as long as the plotted points $(m, \Sigma x_\alpha)$ lie between the lines L_0 and L_1. Inspection is terminated at the first time when the point $(m, \Sigma x_\alpha)$ does not lie between

TABLE 7

m Number of Observations	a_m Acceptance Number	x Observed Value	Σx Cumulated Sum of Observed Values	r_m Rejection Number
1	. . . .	151	151	334
2	139	144	295	476
3	281	121	416	619
4	424	137	553	761
5	566	138	691	904
6	709	136	827	1046
7	851	155	982	1189
8	994	160	1142	1331
9	1136	144	1286	1474
10	1279	145	1431	1616
11	1421	130	1561	1759
12	1564	120	1681	1901
13	1706	104	1785	2044
14	1849	140	1925	2186
15	1991	125	2050	2329
16	2134	106	2156	2471
17	2276	145	2301	2614
18	2419	123	2424	2756
19	2561	138	2562	2899
20	2704	108	2670	3041
21	2846	. . .	. . . .	3184
22	2989	. . .	. . . .	3326
23	3131	. . .	. . . .	3469
24	3274	. . .	. . . .	3611
25	3416	. . .	. . . .	3754

L_0 and L_1. If it lies on L_0 or below the lot is accepted, and if it lies on L_1 or above the lot is rejected.

The common slope of the lines L_0 and L_1 is given by

$$(7{:}14) \qquad s = \frac{\theta_0 + \theta_1}{2}$$

The intercept of L_0 is equal to

$$(7{:}15) \qquad h_0 = \frac{\sigma^2}{\theta_1 - \theta_0} \log \frac{\beta}{1 - \alpha}$$

and the intercept of L_1 is given by

$$(7{:}16) \qquad h_1 = \frac{\sigma^2}{\theta_1 - \theta_0} \log \frac{1 - \beta}{\alpha}$$

7.4   The Operating Characteristic (OC) Curve of the Test

Let $L(\theta)$ denote the probability that the sequential test will lead to the acceptance of the lot when θ is the true mean value. The function $L(\theta)$ is called the operating characteristic function of the test. Approximate formulas for the OC function are derived in Section 3.4 and the general results are applied to testing the mean of a normal population. [See equation (3:48).] It is shown there that

$$(7:17) \qquad L(\theta) \sim \frac{\left(\dfrac{1-\beta}{\alpha}\right)^h - 1}{\left(\dfrac{1-\beta}{\alpha}\right)^h - \left(\dfrac{\beta}{1-\alpha}\right)^h}$$

where

$$(7:18) \qquad h = \frac{\theta_1 + \theta_0 - 2\theta}{\theta_1 - \theta_0}$$

It can be seen from (7:17) and (7:18) that $L(\theta)$ is an increasing function of h and h is a decreasing function of θ. Hence $L(\theta)$ is a decreasing function of θ.

For $\theta = -\infty,\ \theta_0,\ (\theta_0 + \theta_1)/2,\ \theta_1,\ +\infty$ the values of $L(\theta)$ obtained from (7:17) are given as follows.[2]

$$(7:19) \qquad L(-\infty) = 1; \quad L(\theta_0) = 1 - \alpha$$

$$L\left(\frac{\theta_0 + \theta_1}{2}\right) = \frac{\log \dfrac{1-\beta}{\alpha}}{\log \dfrac{1-\beta}{\alpha} - \log \dfrac{\beta}{1-\alpha}}$$

$$L(\theta_1) = \beta$$

$$L(\infty) = 0$$

The computation of these five points of the OC curve will suffice in many applications.

It may be of interest to express $L(\theta)$ in terms of the intercepts h_0

---

[2] For $\theta = \dfrac{\theta_1 + \theta_2}{2}$ we have $h = 0$ and the limiting value of the right-hand member of (7.17) as $h \to 0$ is equal to $\dfrac{\log \dfrac{1-\beta}{\alpha}}{\log \dfrac{1-\beta}{\alpha} - \log \dfrac{\beta}{1-\alpha}}.$

and h_1 and the common slope s of the lines L_0 and L_1.[3] From (7:17) and (7:18) it follows that

$$(7:20) \qquad L(\theta) \sim \frac{e^{\frac{\theta_1+\theta_0-2\theta}{\theta_1-\theta_0}\log\frac{1-\beta}{\alpha}} - 1}{e^{\frac{\theta_1+\theta_0-2\theta}{\theta_1-\theta_0}\log\frac{1-\beta}{\alpha}} - e^{\frac{\theta_1+\theta_0-2\theta}{\theta_1-\theta_0}\log\frac{\beta}{1-\alpha}}}$$

Since $h_0 = \dfrac{\sigma^2}{\theta_1-\theta_0}\log\dfrac{\beta}{1-\alpha}$, $h_1 = \dfrac{\sigma^2}{\theta_1-\theta_0}\log\dfrac{1-\beta}{\alpha}$ and $s = \dfrac{\theta_1+\theta_0}{2}$, we obtain from (7:20)

$$(7:21) \qquad L(\theta) \sim \frac{e^{\frac{2}{\sigma^2}(s-\theta)h_1} - 1}{e^{\frac{2}{\sigma^2}(s-\theta)h_1} - e^{\frac{2}{\sigma^2}(s-\theta)h_0}}$$

7.5 The Average Amount of Inspection Required by the Test

In Section 3.5 the following approximation formula is derived for the expected value $E_\theta(n)$ of the number n of observations required by the sampling plan.

$$(7:22) \qquad E_\theta(n) = \frac{L(\theta)\log\dfrac{\beta}{1-\alpha} + [1 - L(\theta)]\log\dfrac{1-\beta}{\alpha}}{E_\theta(z)}$$

where

$$(7:23) \qquad z = \log\frac{f(x,\theta_1)}{f(x,\theta_0)} = \log\frac{e^{-\frac{1}{2\sigma^2}(x-\theta_1)^2}}{e^{-\frac{1}{2\sigma^2}(x-\theta_0)^2}}$$

$$= \frac{1}{2\sigma^2}[2(\theta_1 - \theta_0)x + \theta_0^2 - \theta_1^2]$$

and $E_\theta(z)$ denotes the expected value of z when θ is the true mean of x. The value of $E_\theta(z)$ is given in Section 3.5, equation (3:60).

$$(7:24) \qquad E_\theta(z) = \frac{1}{2\sigma^2}[2(\theta_1 - \theta_0)\theta + \theta_0^2 - \theta_1^2]$$

Hence

$$(7:25) \qquad E_\theta(n) = 2\sigma^2 \frac{L(\theta)\log\dfrac{\beta}{1-\alpha} + [1 - L(\theta)]\log\dfrac{1-\beta}{\alpha}}{\theta_0^2 - \theta_1^2 + 2(\theta_1 - \theta_0)\theta}$$

$$= \frac{h_1 + L(\theta)(h_0 - h_1)}{\theta - s}$$

[3] See also SRG 255, p. 4.19.

where h_0 and h_1 are the intercepts and s is the common slope of the lines L_0 and L_1.

For $\theta = s$, the right-hand member of (7:25) takes the form 0/0. It is shown in the Appendix, equation (A:99), that the limiting value is given by

$$(7:26) \qquad E_s(n) = \frac{- \log \dfrac{\beta}{1 - \alpha} \log \dfrac{1 - \beta}{\alpha}}{E_s(z^2)}$$

Since $E_s(z) = 0$, $E_s(z^2)$ is equal to the variance $\sigma_z{}^2$ of z. From (7:23) it follows that the variance of z is equal to $(\theta_1 - \theta_0)^2/\sigma^2$. Hence

$$(7:27) \qquad E_s(n) = \frac{- \log \dfrac{\beta}{1 - \alpha} \log \dfrac{1 - \beta}{\alpha}}{(\theta_1 - \theta_0)^2} \sigma^2 = \frac{-h_0 h_1}{\sigma^2}$$

Chapter 8. TESTING THAT THE STANDARD DEVIATION OF A NORMAL DISTRIBUTION DOES NOT EXCEED A GIVEN VALUE

8.1 Formulation of the Problem

Let x be a normally distributed variate. In this section we shall deal with the problem of testing the hypothesis that the standard deviation σ of x does not exceed a given value σ'. There are two cases to be considered: the mean of x is known or unknown. First we shall treat the case when the mean of x is known. If the mean of x is unknown, only a slight modification of the test procedure will be necessary, as will be seen later.

This problem, like the one treated in Section 7, arises frequently in quality control and acceptance inspection. Suppose that x is some measurable quality characteristic of a manufactured product and that x is normally distributed in the population of units produced. Suppose, furthermore, that the quality of the product is considered the better the smaller the standard deviation σ. Thus, there will be, in general, a value σ' such that the product is considered substandard if $\sigma > \sigma'$ and the product is considered satisfactory (meets specification) if $\sigma \leq \sigma'$. Since σ is unknown, the problem is to devise a sampling plan for testing the hypothesis that the product is satisfactory, i.e., that $\sigma \leq \sigma'$.

8.2 Tolerated Risks for Making a Wrong Decision

If the quality of the product is exactly on the margin, i.e., if $\sigma = \sigma'$, it will make no difference whether the product is classified as satisfactory or as substandard. However, if σ is considerably smaller than σ', the classification of the product as substandard will usually be regarded as an error of practical importance. Similarly, if σ is much larger than σ', the classification of the product as satisfactory will be a serious error. Thus, it will be possible to specify two values σ_0 and σ_1 ($\sigma_0 < \sigma'$ and $\sigma_1 > \sigma'$) such that the classification of the product as substandard is considered an error of practical importance whenever $\sigma \leq \sigma_0$, and the classification of the product as satisfactory is regarded as an error of practical consequence whenever $\sigma \geq \sigma_1$; for values σ between σ_0 and σ_1 we do not care particularly which action is taken.

In accordance with the considerations in Section 2.3.2, the risks that we are willing to tolerate may reasonably be stated as follows: The probability of classifying the product as substandard should not exceed a small preassigned value α whenever $\sigma \leqq \sigma_0$, and the probability of classifying the product as satisfactory should not exceed a preassigned value β whenever $\sigma \geqq \sigma_1$.

8.3   The Sequential Probability Ratio Test Corresponding to the Quantities σ_0, σ_1, α, and β

A sampling plan satisfying the requirements regarding the tolerated risks is given by the sequential probability ratio test of strength (α, β) for testing the hypothesis that $\sigma = \sigma_0$ against the alternative that $\sigma = \sigma_1$.

Let $x_1, x_2, \cdots$, etc., denote the successive observations on x. The probability density of the sample $(x_1, \cdots, x_m)$ is given by

$$(8:1) \qquad p_m = \frac{1}{(2\pi)^{\frac{m}{2}} \sigma^m} e^{-\frac{1}{2\sigma^2} \sum_{\alpha=1}^{m} (x_\alpha - \theta)^2}$$

where the value of the mean θ is assumed to be known. Let p_{im} denote the expression we obtain if σ is replaced by σ_i $(i = 0, 1)$ in the right-hand member of (8:1). The sequential probability ratio test is given as follows. The probability ratio p_{1m}/p_{0m} is computed at each stage of the experiment. Additional observations are taken as long as [1]

$$(8:2) \qquad \frac{\beta}{1-\alpha} < \frac{p_{1m}}{p_{0m}} = \frac{\dfrac{1}{\sigma_1{}^m} e^{-\frac{1}{2\sigma_1{}^2} \sum\limits_{\alpha=1}^{m} (x_\alpha - \theta)^2}}{\dfrac{1}{\sigma_0{}^m} e^{-\frac{1}{2\sigma_0{}^2} \sum\limits_{\alpha=1}^{m} (x_\alpha - \theta)^2}} < \frac{1-\beta}{\alpha}$$

The product is classified as satisfactory if

$$(8:3) \qquad \frac{\dfrac{1}{\sigma_1{}^m} e^{-\frac{1}{2\sigma_1{}^2} \sum\limits_{\alpha=1}^{m} (x_\alpha - \theta)^2}}{\dfrac{1}{\sigma_0{}^m} e^{-\frac{1}{2\sigma_0{}^2} \sum\limits_{\alpha=1}^{m} (x_\alpha - \theta)^2}} \leqq \frac{\beta}{1-\alpha}$$

[1] There is a slight approximation involved in the formulas given below, since the constants A and B are put equal to $(1 - \beta)/\alpha$ and $\beta/(1 - \alpha)$ respectively. In this connection see Section 3.3.

The product is classified as substandard if

$$(8:4) \qquad \frac{\dfrac{1}{\sigma_1{}^m} e^{-\frac{1}{2\sigma_1{}^2} \sum\limits_{\alpha=1}^{m} (x_\alpha - \theta)^2}}{\dfrac{1}{\sigma_0{}^m} e^{-\frac{1}{2\sigma_0{}^2} \sum\limits_{\alpha=1}^{m} (x_\alpha - \theta)^2}} \geqq \frac{1 - \beta}{\alpha}$$

Taking logarithms, dividing by $(1/2\sigma_0{}^2) - (1/2\sigma_1{}^2)$ and simplifying, the inequalities (8:2), (8:3), and (8:4) will become

$$(8:5) \qquad \frac{2 \log \dfrac{\beta}{1 - \alpha} + m \log \dfrac{\sigma_1{}^2}{\sigma_0{}^2}}{\dfrac{1}{\sigma_0{}^2} - \dfrac{1}{\sigma_1{}^2}} < \sum_{\alpha=1}^{m} (x_\alpha - \theta)^2 <$$

$$\frac{2 \log \dfrac{1 - \beta}{\alpha} + m \log \dfrac{\sigma_1{}^2}{\sigma_0{}^2}}{\dfrac{1}{\sigma_0{}^2} - \dfrac{1}{\sigma_1{}^2}}$$

$$(8:6) \qquad \sum_{\alpha=1}^{m} (x_\alpha - \theta)^2 \leqq \frac{2 \log \dfrac{\beta}{1 - \alpha} + m \log \dfrac{\sigma_1{}^2}{\sigma_0{}^2}}{\dfrac{1}{\sigma_0{}^2} - \dfrac{1}{\sigma_1{}^2}}$$

and

$$(8:7) \qquad \sum_{\alpha=1}^{m} (x_\alpha - \theta)^2 \geqq \frac{2 \log \dfrac{1 - \beta}{\alpha} + m \log \dfrac{\sigma_1{}^2}{\sigma_0{}^2}}{\dfrac{1}{\sigma_0{}^2} - \dfrac{1}{\sigma_1{}^2}}$$

respectively.

On the basis of the inequalities (8:5), (8:6), and (8:7), the test procedure can be carried out as follows: For each integral value m compute the acceptance number

$$(8:8) \qquad a_m = \frac{2 \log \dfrac{\beta}{1 - \alpha}}{\dfrac{1}{\sigma_0{}^2} - \dfrac{1}{\sigma_1{}^2}} + m \frac{\log \dfrac{\sigma_1{}^2}{\sigma_0{}^2}}{\dfrac{1}{\sigma_0{}^2} - \dfrac{1}{\sigma_1{}^2}}$$

and the rejection number

$$(8{:}9) \qquad r_m = \frac{2 \log \dfrac{1-\beta}{\alpha}}{\dfrac{1}{\sigma_0{}^2} - \dfrac{1}{\sigma_1{}^2}} + m \frac{\log \dfrac{\sigma_1{}^2}{\sigma_0{}^2}}{\dfrac{1}{\sigma_0{}^2} - \dfrac{1}{\sigma_1{}^2}}$$

These acceptance and rejection numbers do not depend on the outcome of the observations and, therefore, they can be computed before inspection starts.   Inspection is continued as long as $a_m < \sum\limits_{\alpha=1}^{m}(x_\alpha - \theta)^2 < r_m$.   The first time that $\Sigma(x_\alpha - \theta)^2$ does not lie between a_m and r_m, inspection is terminated.   If at the final stage $\sum\limits_{\alpha=1}^{m}(x_\alpha - \theta)^2 \leq a_m$ the product is declared satisfactory, and if $\sum\limits_{\alpha=1}^{m}(x_\alpha - \theta)^2 \geq r_m$ the product is declared substandard.

A graphical presentation of the test procedure is shown in Fig. 16.

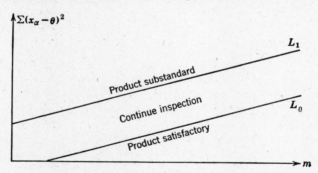

FIG. 16

The number m of observations is measured along the horizontal axis. Since both a_m and r_m are linear functions of m, the points (m, a_m) will lie on a straight line L_0 and the points (m, r_m) will lie on a straight line L_1.   These two lines are parallel and the common slope is given by

$$(8{:}10) \qquad s = \frac{\log \dfrac{\sigma_1{}^2}{\sigma_0{}^2}}{\dfrac{1}{\sigma_0{}^2} - \dfrac{1}{\sigma_1{}^2}}$$

The intercept of L_0 is equal to

$$(8:11) \qquad h_0 = \frac{2 \log \dfrac{\beta}{1 - \alpha}}{\dfrac{1}{\sigma_0{}^2} - \dfrac{1}{\sigma_1{}^2}}$$

and the intercept of L_1 is given by

$$(8:12) \qquad h_1 = \frac{2 \log \dfrac{1 - \beta}{\alpha}}{\dfrac{1}{\sigma_0{}^2} - \dfrac{1}{\sigma_1{}^2}}$$

The lines L_0 and L_1 can be drawn before inspection starts. As inspection goes on the points $[m, \sum\limits_{\alpha=1}^{m} (x_\alpha - \theta)^2]$ are plotted. The first time that the point $[m, \Sigma(x_\alpha - \theta)^2]$ does not lie between the lines L_0 and L_1, inspection is terminated. If the point $[m, \Sigma(x_\alpha - \theta)^2]$ lies on L_0 or below, the hypothesis that the product is satisfactory is accepted; and if the point $[m, \Sigma(x_\alpha - \theta)^2]$ lies on L_1 or above, the product is declared substandard.

8.4   The Operating Characteristic (OC) Function of the Test

For any value σ, let $L(\sigma)$ denote the probability that the test will terminate with the acceptance of the hypothesis that the product is satisfactory. The function $L(\sigma)$ is called the operating characteristic function of the test.

In Section 3.4 a general method is given for deriving an approximation formula for the OC function for any sequential probability ratio test. Applying the result of that section, we obtain

$$(8:13) \qquad L(\sigma) = \frac{\left(\dfrac{1 - \beta}{\alpha}\right)^h - 1}{\left(\dfrac{1 - \beta}{\alpha}\right)^h - \left(\dfrac{\beta}{1 - \alpha}\right)^h}$$

where h is the root of the equation

$$(8:14) \qquad \frac{1}{\sqrt{2\pi}} \frac{\sigma_0{}^h}{\sigma_1{}^h} \int_{-\infty}^{+\infty} \left(\frac{e^{-\frac{1}{2\sigma_1{}^2}(x-\theta)^2}}{e^{-\frac{1}{2\sigma_0{}^2}(x-\theta)^2}}\right)^h e^{-\frac{1}{2\sigma^2}(x-\theta)^2} \, dx = 1$$

It can be seen that the integral on the left side of (8:14) has a finite value only if $(h/\sigma_1{}^2) - (h/\sigma_0{}^2) + (1/\sigma^2) > 0$. In this case, as can be verified, we have

$$(8:15) \qquad \int_{-\infty}^{+\infty} \left(\frac{e^{-\frac{1}{2\sigma_1{}^2}(x-\theta)^2}}{e^{-\frac{1}{2\sigma_0{}^2}(x-\theta)^2}} \right)^h e^{-\frac{1}{2\sigma^2}(x-\theta)^2} dx = \sqrt{\frac{2\pi}{\dfrac{h}{\sigma_1{}^2} - \dfrac{h}{\sigma_0{}^2} + \dfrac{1}{\sigma^2}}}$$

Hence equation (8:14) can be written as

$$(8:16) \qquad \sigma \left(\frac{\sigma_1}{\sigma_0} \right)^h = \sqrt{\frac{1}{\dfrac{h}{\sigma_1{}^2} - \dfrac{h}{\sigma_0{}^2} + \dfrac{1}{\sigma^2}}}$$

Instead of solving (8:16) with respect to h, we shall solve it with respect to σ. We obtain

$$(8:17) \qquad \sigma = \sqrt{\frac{\left(\dfrac{\sigma_0}{\sigma_1} \right)^{2h} - 1}{\dfrac{h}{\sigma_1{}^2} - \dfrac{h}{\sigma_0{}^2}}}$$

With the use of equations (8:13) and (8:17), the OC curve can be plotted as follows. For any given value of h we compute σ and $L(\sigma)$ from equations (8:13) and (8:17). The pair $[\sigma, L(\sigma)]$ obtained in this way gives us a point on the OC curve. Computing $[\sigma, L(\sigma)]$ for a sufficiently large number of values of h, we obtain enough points to draw the OC curve.

For computational purposes, it may be convenient to put [2]

$$(8:18) \qquad \frac{h}{2\sigma_1{}^2} - \frac{h}{2\sigma_0{}^2} = t \quad \text{or} \quad h = \frac{-2t}{\left(\dfrac{1}{\sigma_0{}^2} - \dfrac{1}{\sigma_1{}^2} \right)}$$

Then equations (8:13) and (8:17) can be written as

$$(8:19) \qquad L(\sigma) = \frac{e^{\left(\log \frac{1-\beta}{\alpha} \right)\left(\frac{-2t}{\frac{1}{\sigma_0{}^2} - \frac{1}{\sigma_1{}^2}} \right)} - 1}{e^{\left(\log \frac{1-\beta}{\alpha} \right)\left(\frac{-2t}{\frac{1}{\sigma_0{}^2} - \frac{1}{\sigma_1{}^2}} \right)} - e^{\left(\log \frac{\beta}{1-\alpha} \right)\left(\frac{-2t}{\frac{1}{\sigma_0{}^2} - \frac{1}{\sigma_1{}^2}} \right)}}$$

$$= \frac{e^{-th_1} - 1}{e^{-th_1} - e^{-th_0}}$$

[2] A similar simplification was made by the Statistical Research Group. See SRG 255, p. 6.31. The parameter t used there corresponds to $-t$ here.

and

$$(8{:}20) \qquad \sigma = \sqrt{\frac{e^{\left(\log \frac{\sigma_0{}^2}{\sigma_1{}^2}\right)\left(\frac{-2t}{\frac{1}{\sigma_0{}^2} - \frac{1}{\sigma_1{}^2}}\right)} - 1}{2t}} = \sqrt{\frac{e^{2st} - 1}{2t}}$$

where s is the common slope and h_0 and h_1 are the intercepts of the lines L_0 and L_1. Equations (8:19) and (8:20) may be more convenient for the computation of the OC curve than the original equations (8:13) and (8:17).

For $\sigma = 0$, σ_0, $\sqrt{s}$, σ_1, $+\infty$ the values of $L(\sigma)$ are given as follows:

$$(8{:}21) \qquad\qquad L(0) = 1$$

$$L(\sigma_0) = 1 - \alpha$$

$$L(\sqrt{s}) = \frac{h_1}{h_1 - h_0}$$

$$L(\sigma_1) = \beta$$

$$L(\infty) = 0$$

These five points already determine roughly the shape of the OC curve and in many instances it will not be necessary to compute further points.

8.5 The Average Amount of Inspection Required by the Test

According to the results in Section 3.5, an approximation formula for the expected value $E_\sigma(n)$ of the number n of observations required by the sampling plan is given by

$$(8{:}22) \qquad E_\sigma(n) = \frac{L(\sigma) \log \dfrac{\beta}{1 - \alpha} + [1 - L(\sigma)] \log \dfrac{1 - \beta}{\alpha}}{E_\sigma(z)}$$

where

$$(8{:}23) \quad z = \log \frac{\dfrac{1}{\sigma_1} e^{-\frac{1}{2\sigma_1{}^2}(x-\theta)^2}}{\dfrac{1}{\sigma_0} e^{-\frac{1}{2\sigma_0{}^2}(x-\theta)^2}} = \log \frac{\sigma_0}{\sigma_1} + \frac{1}{2}\left(\frac{1}{\sigma_0{}^2} - \frac{1}{\sigma_1{}^2}\right)(x - \theta)^2$$

and $E_\sigma(z)$ denotes the expected value of z when σ is the standard deviation of x.   We have

$$(8{:}24) \qquad E_\sigma(z) = \frac{1}{2}\left(\frac{1}{\sigma_0{}^2} - \frac{1}{\sigma_1{}^2}\right) E(x - \theta)^2 + \log \frac{\sigma_0}{\sigma_1}$$

$$= \frac{1}{2}\left(\frac{1}{\sigma_0{}^2} - \frac{1}{\sigma_1{}^2}\right) \sigma^2 + \log \frac{\sigma_0}{\sigma_1}$$

Hence, substituting the right-hand member of (8:24) for $E_\sigma(z)$ in (8:22) we obtain [3]

$$(8{:}25) \quad E_\sigma(n) = \frac{L(\sigma)\left[\log \dfrac{\beta}{1 - \alpha} - \log \dfrac{1 - \beta}{\alpha}\right] + \log \dfrac{1 - \beta}{\alpha}}{\dfrac{1}{2}\left(\dfrac{1}{\sigma_0{}^2} - \dfrac{1}{\sigma_1{}^2}\right)\sigma^2 + \log \dfrac{\sigma_0}{\sigma_1}}$$

$$= \frac{L(\sigma)(h_0 - h_1) + h_1}{\sigma^2 - s}$$

For $\sigma = \sqrt{s}$ the expected value of z is equal to 0 and the right-hand member of (8:25) takes the form $0/0$.   According to equation (A:99) in the Appendix, the limiting value is given by

$$(8{:}26) \qquad E_{\sqrt{s}}(n) = \frac{- \log \dfrac{\beta}{1 - \alpha}\, \log \dfrac{1 - \beta}{\alpha}}{E_{\sqrt{s}}(z^2)}$$

Since $E_{\sqrt{s}}(z) = 0$, $E_{\sqrt{s}}(z^2)$ is equal to the variance of z when $\sigma = \sqrt{s}$.   It follows easily from (8:23) that this variance is equal to $\dfrac{1}{2}\left(\dfrac{1}{\sigma_0{}^2} - \dfrac{1}{\sigma_1{}^2}\right)^2 s^2$.   Hence

$$(8{:}27) \qquad E_{\sqrt{s}}(n) = \frac{- \log \dfrac{\beta}{1 - \alpha}\, \log \dfrac{1 - \beta}{\alpha}}{\dfrac{1}{2}\left(\dfrac{1}{\sigma_0{}^2} - \dfrac{1}{\sigma_1{}^2}\right)^2 s^2} = \frac{-h_0 h_1}{2s^2}$$

[3] The expression of $E_\sigma(n)$ in terms of the slope and intercepts of the decision lines is contained also in SRG 255, p. 6.34.

8.6   Modification of the Test Procedure When the Population Mean Is Not Known [4]

If the mean θ of x is not known, the following two modifications of the test procedure are to be made: (1) replace $\sum_{\alpha=1}^{m} (x_\alpha - \theta)^2$ by

$\sum_{\alpha=1}^{m} (x_\alpha - \bar{x})$ where $\bar{x} = (x_1 + \cdots + x_m)/m$; (2) the acceptance number a_m is replaced by a_{m-1} and the rejection number r_m is replaced by r_{m-1}. Thus, if the mean is unknown, the acceptance and rejection numbers at the mth trial are equal to the acceptance and rejection numbers corresponding to the $(m-1)$th trial when the mean is known.

The formula for the OC curve remains unchanged and the expected value of the number of observations required by the test is larger by 1 when the mean is unknown than when the mean is known.

[4] The result contained in this section was found by C. Stein and M. A. Girshick, independently of each other. The proof is based on a transformation of the observations which reduces this case to the case when the mean is known. See Girshick's paper, "Contribution to the Theory of Sequential Analysis," *The Annals of Mathematical Statistics*, June, 1946.

Chapter 9. TESTING THAT THE MEAN OF A NORMAL DIS-TRIBUTION WITH KNOWN VARIANCE IS EQUAL TO A SPECIFIED VALUE

9.1 Formulation of the Problem

Let x be a quality characteristic of a product, such as weight, diameter, or hardness. Suppose that x is normally distributed in the population of all units produced and that the standard deviation σ of x is known but the mean θ of x is unknown. Suppose, furthermore, that a particular value of θ, say θ_0, is considered the most desirable value for the product. In general, the greater the absolute deviation of the true value θ from the most desirable value θ_0, the less satisfactory the product. Since the manufacturer would like to achieve and maintain the value θ_0 of θ as closely as possible, he will be interested in testing the hypothesis that $\theta = \theta_0$. If the evidence supplied by a sample should indicate that $\theta \neq \theta_0$, he will try to improve the production process. Of course, if $\theta \neq \theta_0$ but is near θ_0, there is no particular need to improve the production, and acceptance of the hypothesis that $\theta = \theta_0$ would not be a serious error. However, there will be, in general, a positive value δ such that the acceptance of the hypothesis that $\theta = \theta_0$ is regarded as an error of practical importance whenever $\left| \dfrac{\theta - \theta_0}{\sigma} \right| \geqq \delta$.

The situation described in the preceding paragraph will thus lead to the following problem: A sampling plan is to be devised for which the probability that the hypothesis that $\theta = \theta_0$ will be rejected (the product will be declared substandard) does not exceed a small pre-assigned value α when $\theta = \theta_0$, and the probability of accepting the hypothesis that $\theta = \theta_0$ (declaring the product satisfactory) does not exceed a small preassigned value β whenever $\left| \dfrac{\theta - \theta_0}{\sigma} \right| \geqq \delta$.

9.2 A Sequential Sampling Plan Satisfying the Imposed Requirements

It has been shown in Section 4.1.4 that an adequate sampling plan for the problem described in Section 9.1 is given as follows. Compute the ratio

$$(9{:}1) \qquad \frac{p_{1m}}{p_{0m}} = \frac{1}{2} \frac{e^{-\frac{1}{2\sigma^2}\sum\limits_{\alpha=1}^{m}(x_\alpha-\theta_0-\delta\sigma)^2} + e^{-\frac{1}{2\sigma^2}\sum\limits_{\alpha=1}^{m}(x_\alpha-\theta_0+\delta\sigma)^2}}{e^{-\frac{1}{2\sigma^2}\sum\limits_{\alpha=1}^{m}(x_\alpha-\theta_0)^2}}$$

at each stage of the experiment. Continue taking observations as long as

$$(9{:}2) \qquad B < \frac{p_{1m}}{p_{0m}} < A$$

Accept the hypothesis that the product is satisfactory if

$$(9{:}3) \qquad \frac{p_{1m}}{p_{0m}} \leqq B$$

Reject the hypothesis that the product is satisfactory if

$$(9{:}4) \qquad \frac{p_{1m}}{p_{0m}} \geqq A$$

To satisfy the requirements imposed regarding the probabilities of making wrong decisions, for all practical purposes we may put $A = (1-\beta)/\alpha$ and $B = \beta/(1-\alpha)$.

The expression for p_{1m}/p_{0m} given in (9:1) can be simplified to

$$(9{:}5) \qquad \frac{p_{1m}}{p_{0m}} = \frac{1}{2} e^{-\frac{1}{2}m\delta^2}(e^{\frac{\delta}{\sigma}\Sigma(x_\alpha-\theta_0)} + e^{-\frac{\delta}{\sigma}\Sigma(x_\alpha-\theta_0)})$$

$$= e^{-\frac{1}{2}m\delta^2} \cosh\left[\frac{\delta}{\sigma}\sum_{\alpha=1}^{m}(x_\alpha-\theta_0)\right]$$

Substituting this value of p_{1m}/p_{0m} in (9:2), (9:3), and (9:4) and taking logarithms, we find that these inequalities become

$$(9{:}6) \quad \log B + m\frac{\delta^2}{2} < \log \cosh\left[\frac{\delta}{\sigma}\sum_{\alpha=1}^{m}(x_\alpha-\theta_0)\right] < \log A + m\frac{\delta^2}{2}$$

$$(9{:}7) \qquad \log \cosh\left[\frac{\delta}{\sigma}\Sigma(x_\alpha-\theta_0)\right] \leqq \log B + m\frac{\delta^2}{2}$$

and

$$(9{:}8) \qquad \log \cosh\left[\frac{\delta}{\sigma}\Sigma(x_\alpha-\theta_0)\right] \geqq \log A + m\frac{\delta^2}{2}$$

With the use of inequalities (9:6), (9:7), and (9:8), the test procedure is carried out as follows. At each stage of the experiment we compute $Z_m = \log \cosh \left[\dfrac{\delta}{\sigma} \sum_{\alpha=1}^{m}(x_\alpha - \theta_0) \right]$. The first time that Z_m does not lie between $\log B + [m(\delta^2/2)]$ and $\log A + [m(\delta^2/2)]$ we terminate the process. The hypothesis that $\theta = \theta_0$ is accepted if $Z_m \leqq \log B + [m(\delta^2/2)]$, and rejected if $Z_m \geqq \log A + [m(\delta^2/2)]$.

The computation of Z_m at each stage of the experiment is somewhat cumbersome. However, if $\left| \dfrac{\delta}{\sigma} \Sigma(x_\alpha - \theta_0) \right|$ is greater than 3, $Z_m = \log \cosh \left| \dfrac{\delta}{\sigma} \Sigma(x_\alpha - \theta_0) \right|$ is very nearly equal to $\left| \dfrac{\delta}{\sigma} \Sigma(x_\alpha - \theta_0) \right| - \log 2$.[1] When this approximation to Z_m is used, inequalities (9:6), (9:7), and (9:8) simplify to

$$(9:9) \quad \frac{\sigma}{\delta}(\log B + \log 2) + m \frac{\sigma\delta}{2} < \left| \Sigma(x_\alpha - \theta_0) \right| <$$
$$\frac{\sigma}{\delta}(\log A + \log 2) + m \frac{\sigma\delta}{2}$$

$$(9:10) \quad \left| \Sigma(x_\alpha - \theta_0) \right| \leqq \frac{\sigma}{\delta}(\log B + \log 2) + m \frac{\sigma\delta}{2}$$

and

$$(9:11) \quad \left| \Sigma(x - \theta_0) \right| \geqq \frac{\sigma}{\delta}(\log A + \log 2) + m \frac{\sigma\delta}{2}$$

respectively. For all practical purposes inequalities (9:9), (9:10), and (9:11) may be used instead of (9:6), (9:7), and (9:8) whenever $\dfrac{\delta}{\sigma} \left| \Sigma(x_\alpha - \theta_0) \right| \geqq 3$.

The following is an alternative computational procedure which may be found useful. Consider the equation in u.

$$(9:12) \quad \log \cosh \left| u \right| = v$$

This has exactly one positive solution if $v \geqq 0$. The root of this equation is given by

$$(9:13) \quad \left| u \right| = \phi(v) = \log \left(e^v + \sqrt{e^{2v} - 1} \right)$$

[1] See also SRG 255, p. B.15.

The function $\phi(v)$ can easily be tabulated. In terms of the function $\phi(v)$, inequalities (9:6), (9:7), and (9:8) can be written as

$$(9:14) \quad \frac{\sigma}{\delta}\phi\left(\log B + m\,\frac{\delta^2}{2}\right) < \left|\,\Sigma(x_\alpha - \theta_0)\,\right| < \frac{\sigma}{\delta}\phi\left(\log A + m\,\frac{\delta^2}{2}\right)$$

$$(9:15) \quad \left|\,\Sigma(x_\alpha - \theta_0)\,\right| \leq \frac{\sigma}{\delta}\phi\left(\log B + m\,\frac{\delta^2}{2}\right)$$

and

$$(9:16) \quad \left|\,\Sigma(x_\alpha - \theta_0)\,\right| \geq \frac{\sigma}{\delta}\phi\left(\log A + m\,\frac{\delta^2}{2}\right)$$

When inequalities (9:14), (9:15), and (9:16) are used, the test can be carried out as follows. For each integral value m we compute the acceptance number

$$(9:17) \quad a_m = \frac{\sigma}{\delta}\phi\left(\log B + m\,\frac{\delta^2}{2}\right)$$

and the rejection number

$$(9:18) \quad r_m = \frac{\sigma}{\delta}\phi\left(\log A + m\,\frac{\delta^2}{2}\right)$$

These acceptance and rejection numbers can be computed before experimentation starts. Additional observations are taken as long as $a_m < \left|\,\Sigma(x_\alpha - \theta_0)\,\right| < r_m$. If $\left|\,\Sigma(x_\alpha - \theta_0)\,\right| \leq a_m$ the hypothesis that $\theta = \theta_0$ is accepted and if $\left|\,\Sigma(x_\alpha - \theta_0)\,\right| \geq r_m$ the hypothesis that $\theta = \theta_0$ is rejected.

PART III. THE PROBLEM OF MULTI-VALUED DECISIONS AND ESTIMATION

Chapter 10. THE CHOICE OF A HYPOTHESIS FROM A SET OF MUTUALLY EXCLUSIVE HYPOTHESES (MULTI-VALUED DECISION)

10.1 Formulation of the Problem

Part I has been devoted exclusively to the discussion of the problem of testing a statistical hypothesis. In such problems only one of two possible decisions can be made: the hypothesis is either rejected or accepted. Thus, we can say that testing a hypothesis is a two-valued decision problem, since the decision can take only the two values: acceptance and rejection. Let $\bar{H}$ denote the negation of the hypothesis H to be tested. Then testing the hypothesis H is the same as choosing between the two competing hypotheses H and $\bar{H}$.

It has been pointed out in Section 1.3.5 that testing a hypothesis H arises frequently as a consequence of the problem of deciding between two alternative courses of action, say action 1 and action 2. Suppose that the preference for one or the other action depends on the value of an unknown parameter θ of the distribution of a random variable x. Let ω denote the set of all values of θ for which action 1 is preferred to action 2 (or at least not less desirable than action 2). If a decision is to be made on the basis of a finite number of observations on x, this leads to the problem of testing the hypothesis H that the true value θ lies in ω. If H is accepted, we decide for action 1, and if H is rejected we decide for action 2. In applications it happens frequently that there are more than two alternative courses of action, one of which is to be chosen. Suppose that there are k $(k > 2)$ alternative actions, say action 1, action 2, $\cdots$, action k, and that one of them is to be chosen on the basis of some observations on the random variable x. Suppose, furthermore, that the relative degree of preference for these actions depends on the value of a parameter θ of the distribution of x. Then it will be possible, in general, to subdivide the totality of all possible values of θ into k mutually exclusive parts ω_1, ω_2, $\cdots$, ω_k such that action j is preferable to all other actions $i \neq j$ if, and only if, the true value θ lies in ω_j. Let H_j denote the hypothesis that θ lies in ω_j $(j = 1, \cdots, k)$. Then the problem of deciding for a particular action reduces to the problem of choosing one of the hypotheses $H_1, \cdots, H_k$. If H_i is accepted we decide to take action i. Such a problem may be called a multi-valued decision problem, since the decision to be made can take k values: We may accept H_1, or H_2, $\cdots$, or H_k.

In this section we shall deal with the problem of choosing one out of k mutually exclusive and exhaustive hypotheses, $H_1, \cdots, H_k$, on the basis of some observations on the random variable x under consideration.[1] The problem of testing a hypothesis is contained in this as a special case when $k = 2$.

The following simple example may serve as an illustration. Suppose that x is a measurable quality characteristic of a product which is normally distributed in the population of units produced. Suppose, furthermore, that the quality of the product is regarded the better the higher the mean value θ of x. Assume that the following three alternative actions are under consideration by the manufacturer: (1) to sell the product at the regular market price, (2) to label the product as second rate quality and sell it at a reduced price, (3) to withhold the product from the market. Let a and b $(a < b)$ be two values of θ such that the manufacturer prefers action 3 if $\theta \leq a$, he prefers action 2 if $a < \theta < b$, and he prefers action 1 if $\theta \geq b$. Let H_1 denote the hypothesis that $\theta \leq a$, H_2 the hypothesis that $a < \theta < b$, and H_3 the hypothesis that $\theta \geq b$. If the value of θ is unknown and if the manufacturer must decide which action should be taken on the basis of some observations on x, he is faced with the multi-valued decision problem of choosing one of the mutually exclusive hypotheses H_1, H_2, and H_3.

10.2   The General Nature of a Sequential Sampling Plan for Selecting a Hypothesis from a Set of Mutually Exclusive Hypotheses

A sequential sampling plan for choosing one of k mutually exclusive and exhaustive hypotheses $H_1, \cdots, H_k$ may be described as follows. A rule is given for making one of the following $(k + 1)$ decisions at each stage of the experiment (at the mth trial for each integral value of m): (1) to terminate experimentation with the acceptance of H_1; (2) to terminate experimentation with the acceptance of H_2; $\cdots$; (k)

---

[1] This problem in the non-sequential case, that is, when the total number of observations to be made is determined in advance, has been treated in several previous publications. See, for example, the author's article "Statistical Decision Functions Which Minimize the Maximum Risk," *The Annals of Mathematics*, April, 1945.

to terminate experimentation with the acceptance of H_k; $(k + 1)$ to continue the experiment by making an additional observation. Such a procedure is carried out sequentially. On the basis of the first observation one of the aforementioned $(k + 1)$ decisions is made. If one of the first k decisions is made, the process is terminated. If the last decision is made, a second trial is performed. Again, on the basis of the first two observations, one of the $(k + 1)$ decisions is made. If the last decision is made, a third trial is performed, and so on. The process is continued until one of the first k decisions is made.

In more precise mathematical terms, a sequential sampling plan may be described as follows. Let R_m denote the totality of all possible samples of size m, i.e., R_m is the m-dimensional sample space. For each positive integral value of m, the m-dimensional sample space is split into $(k + 1)$ mutually exclusive parts, R_{m1}, R_{m2}, $\cdots$, R_{mk} and $R_{m,k+1}$. If the first observation x_1 lies in R_{1i} where $i \leq k$, the process is terminated with the acceptance of H_i. If x_1 lies in $R_{1,k+1}$ a second observation x_2 is made. Again, if (x_1, x_2) lies in some R_{2i} with $i \leq k$, the process is terminated with the acceptance of H_i. If (x_1, x_2) lies in $R_{2,k+1}$ a third trial is performed, and so on. This process is stopped at the first time when the sample $(x_1, \cdots, x_m)$ lies in R_{mi} for some value $i \leq k$. Thus, a sequential sampling plan is completely defined by the sets R_{m1}, $\cdots$, $R_{m,k+1}$. Since these sets are mutually exclusive and add up to the whole sample space R_m, it is sufficient to define any k of these sets, since they determine uniquely the remaining set.

For any m, the subdivision of the sample space R_m into the $(k + 1)$ parts R_{m1}, $\cdots$, $R_{m,k+1}$ can be made in many ways, and a fundamental problem is that of a proper choice of these sets. In order to set up principles for this choice, in the next section we shall study the consequences of any particular choice.

10.3 Consequences of the Choice of Any Particular Sequential Sampling Plan

After a particular choice of the sets R_{m1}, $\cdots$, $R_{m,k+1}$ has been made, i.e., a particular sequential sampling plan has been adopted, for any $i \leq k$ the probability that the process will terminate with the acceptance of H_i depends only on the distribution of the random variable x under consideration. Since it is assumed that the distribution of x is known except for the values of a finite number of parameters θ_1, $\cdots$, θ_r, the probability that H_i will be accepted will be a function of these parameters. To simplify notation, we shall use the letter θ without subscript to denote the set of all r parameters $\theta_1, \cdots, \theta_r$. Let $L_i(\theta)$ denote the probability that the adopted sequential sampling plan will terminate with the acceptance of H_i $(i = 1, \cdots, k)$. We shall refer to the set of functions $L_1(\theta), L_2(\theta), \cdots, L_k(\theta)$ as the operating characteristics of the sampling plan. We shall consider only sampling plans for which the probability is 1 that the process will eventually terminate. Then we have

$$(10:1) \qquad\qquad L_1(\theta) + \cdots + L_k(\theta) = 1$$

and, therefore, one of the functions $L_1(\theta), \cdots, L_k(\theta)$ is determined by the other $k - 1$.

The operating characteristics represent the accomplishment of the sampling plan in giving protection against possible wrong decisions. For any parameter point θ, the probability of accepting the correct hypothesis, i.e., the hypothesis which is consistent with parameter point θ, can be obtained immediately from the operating characteristics. Since the hypotheses $H_1, \cdots, H_k$ are mutually exclusive and exhaustive, for any given parameter point θ one, and only one, of the hypotheses $H_1, \cdots, H_k$ will be consistent with a given θ. If H_i is the hypothesis consistent with a given θ, the probability of making a correct decision when this θ is true is equal to $L_i(\theta)$. The operating characteristics of a sampling plan are considered the more favorable the higher the probability for making correct decisions for the various possible parameter points θ.

The price we have to pay for the accomplishment of the sampling plan in giving protection against wrong decisions is represented by the number n of observations required by the sampling plan. Since n is a random variable, we shall consider, as in testing a hypothesis, the expected value of n. After a particular sampling plan has been adopted, the expected value of n will be a function of the parameter point θ only. As in testing hypotheses, we shall denote the expected value of n, when θ is true, by $E_\theta(n)$, and we shall refer to $E_\theta(n)$ as the average sample number (ASN) function of the sampling plan.

In conclusion we may say that the most important consequences of any particular choice of a sampling plan are given by the operating characteristics and the ASN function of the adopted sampling plan. The operating characteristics represent the accomplishments of the sampling plan and the ASN function represents the price paid for these accomplishments.

10.4   Principles for the Selection of a Sequential Sampling Plan

10.4.1   Dependence of Importance of Possible Wrong Decisions on the Parameter Point θ

To set up principles for the selection of a sequential sampling plan it will be necessary to investigate the dependence of the importance of possible wrong decisions on the parameter point. Let ω_i denote the set of parameter points θ consistent with H_i ($i = 1, \cdots, k$), i.e., H_i is precisely the statement that the true parameter point θ is included in ω_i. If the true θ is in ω_i but not far from ω_j for some $j \neq i$, the acceptance of H_j will not be regarded, in general, as a serious error. However, if θ is far from ω_j and H_j is accepted, the error committed will usually be of considerable practical consequence.

As an illustration, consider again the example given in Section 10.1. The decision to withhold the product from the market will be considered an error of little practical significance if θ is only slightly above a. The seriousness of this error will, however, increase with increasing value of θ. If θ is substantially above a, the decision to withhold the product will be regarded as an error of considerable practical importance. Similarly, the decision to try to sell the product at regular market price will not be a serious error if θ is just slightly below b, but the importance of this error will increase with decreasing value of θ.

It will frequently be possible to express the importance of the various possible wrong decisions by k functions $w_1(\theta), \cdots, w_k(\theta)$, where $w_j(\theta)$ is a non-negative function expressing the importance of the error committed by accepting H_j when θ is true. In industrial problems, $w_j(\theta)$ may be thought of as expressing the financial loss caused by taking the action corresponding to the acceptance of H_j when θ is true. We shall, of course, put $w_j(\theta) = 0$ for all points θ in ω_j, since for such points θ the acceptance of H_j is a correct decision. We shall refer to the functions $w_1(\theta), \cdots, w_k(\theta)$ as error weight functions, or more briefly as weight functions.

The choice of a sampling plan will be influenced by the weight functions $w_1(\theta), \cdots, w_k(\theta)$. The determination of these weight functions cannot be regarded as a statistical problem. They will be chosen on the basis of practical considerations in each particular problem.

10.4.2   The Risk Function Associated with a Given Sampling Plan

For any parameter point θ we shall mean by the risk $r(\theta)$ the expected value of the loss caused by possible wrong decisions when θ is true. Since the probability of accepting H_i is equal to $L_i(\theta)$ and since the loss caused by this decision is given by $w_i(\theta)$, the expected value of the loss is equal to

$$(10:2) \quad r(\theta) = L_1(\theta)w_1(\theta) + L_2(\theta)w_2(\theta) + \cdots + L_k(\theta)w_k(\theta)$$

We shall refer to $r(\theta)$ as the risk function of the sampling plan.[2]

We shall judge the relative merits of a sampling plan by its risk function $r(\theta)$ and ASN function $E_\theta(n)$.

10.4.3   The Risk Function and the ASN Function as a Basis for the Selection of a Sequential Sampling Plan

A sequential sampling plan is the better the smaller the risk $r(\theta)$ and the smaller the expected value $E_\theta(n)$ of the number of observations. These two desiderata of a sampling plan are somewhat in conflict, since the smaller we make $r(\theta)$, the larger, in general, will be the number of observations required by the plan. To achieve a reasonable compromise between these two conflicting desiderata, one may proceed as follows. First we impose the condition that the risk $r(\theta)$ shall not exceed a certain prescribed positive value r_0, i.e.,

$$(10:3) \quad\quad\quad\quad\quad\quad\quad r(\theta) \leq r_0$$

for all parameter points θ. We then consider only sampling plans for which (10:3) is fulfilled. From this class of sampling plans we try to select one for which $E_\theta(n)$ is as small as possible.

To impose first the condition (10:3) and then to try to minimize with respect to the expected number of observations does not seem to be an unreasonable procedure, since the risk function $r(\theta)$ is perhaps of primary importance.[3]

The choice of the upper limit r_0 of the risk is not a statistical problem. It will be determined on the basis of practical considerations in each particular case.

---

[2] Another possible definition of the risk function could be given by including also the expected value of the cost of experimentation. If c denotes the cost of taking a single observation, the expected value of the cost of experimentation is equal to $cE_\theta(n)$ and the risk is given by

$$(10:2^*) \quad\quad\quad\quad r^*(\theta) = \sum_{i=1}^{k} L_i(\theta)w_i(\theta) + cE_\theta(n)$$

If the cost of experimentation is not proportional with the number of observations, but is given by the cost function $c(n)$, then the term $cE_\theta(n)$ in (10:2*) is to be replaced by $E_\theta[c(n)]$.

[3] Using the risk function $r^*(\theta)$, as given in (10:2*), a sampling plan for which the maximum value of $r^*(\theta)$ with respect to θ is minimized may be regarded as an optimum plan. If this definition of an optimum sampling plan is accepted, no condition of the type (10:3) is imposed; we simply try to find a plan for which the maximum of $r^*(\theta)$ with respect to θ takes the smallest possible value.

10.4.4 The Use of Certain Simple Weight Functions

The construction of specific weight functions $w_1(\theta)$, $\cdots$, $w_k(\theta)$ in a given problem may occasionally run into practical difficulties. Although in industrial problems $w_j(\theta)$ could be assumed to be equal to the financial loss (or estimated financial loss) caused by the acceptance of H_j when θ is true, in purely scientific investigations it is rather difficult to give a reasonable measure of the loss caused by accepting a wrong hypothesis.

Even if the difficulties in measuring the loss caused by possible wrong decisions are disregarded, we still face the practical difficulty that the weight functions $w_1(\theta)$, $\cdots$, $w_k(\theta)$ in a given problem may be too involved to be manageable. Thus, there is a need for simplification.

The choice of the sampling plan is usually not very dependent on the exact shape of the weight functions. It will, therefore, be frequently satisfactory to use some rough approximations, reproducing only the main features of the weight functions. A very rough, but for many applications satisfactory, approximation can be obtained by replacing $w_j(\theta)$ by $\bar{w}_j(\theta)$ defined as follows:

(10:4) $\bar{w}_j(\theta) = 0$ if $w_j(\theta)$ is less than or equal to a certain value c_j

$\quad\quad = c$ if $w_j(\theta) > c_j$

where c is some positive constant. Thus, $\bar{w}_j(\theta)$ can take only two values, 0 and c. There is no loss of generality in putting $c = 1$, since this can be achieved by multiplication by a proportionality factor which has no effect on the selection of the sampling plan.

In what follows in this and the following section, we shall consider only the weight functions $\bar{w}_j(\theta)$. We shall call the set of all parameter points θ for which $\bar{w}_i(\theta) = 0$ and $\bar{w}_j(\theta) = 1$ for $j \neq i$ the zone of preference for acceptance of H_i. The set of points θ for which $\bar{w}_i(\theta) = \bar{w}_j(\theta) = 0$ and $\bar{w}_k(\theta) = 1$ for $k \neq i, j$ will be called the zone of indifference between H_i and H_j. Similarly, the set of points θ for which $\bar{w}_i(\theta) = \bar{w}_j(\theta) = \bar{w}_m(\theta) = 0$ and $\bar{w}_l(\theta) = 1$ for $l \neq i, j, m$ will be called the zone of indifference among the hypotheses H_i, H_j, and H_m, and so on.

If we deal with the problem of testing a hypothesis H, then $k = 2$, $H_1 = H$, and H_2 is equal to the negation $\bar{H}$ of H. The zone of preference for acceptance of H, the zone of preference for acceptance of $\bar{H}$, and the zone of indifference between H and $\bar{H}$ defined here correspond to the zone of preference for acceptance, zone of preference for rejection, and zone of indifference discussed in Section 2.3.1.

To illustrate the meaning of the various zones defined here, we consider again the example discussed in Section 10.1. In this example H_1 is the hypothesis that $\theta \leqq a$, H_2 is the hypothesis that $a < \theta < b$, and H_3 is the hypothesis that $\theta \geqq b$. The functions $\bar{w}_1(\theta)$, $\bar{w}_2(\theta)$, and $\bar{w}_3(\theta)$ may reasonably be defined as follows:

$$\bar{w}_1(\theta) = 0 \quad \text{for } \theta < a + \Delta$$

$$= 1 \quad \text{for } \theta \geqq a + \Delta \text{ where } \Delta \text{ is a certain positive quantity}$$

$$\bar{w}_2(\theta) = 0 \quad \text{if } a - \Delta < \theta < b + \Delta \text{ and } = 1 \text{ elsewhere}$$

$$\bar{w}_3(\theta) = 0 \quad \text{if } \theta \geqq b - \Delta \text{ and } = 1 \text{ elsewhere}$$

Then the zone of preference for acceptance of H_1 is the set of values of θ for which $\theta \leqq a - \Delta$. The zone of preference for acceptance of H_2 is given by the inequality $a + \Delta \leqq \theta < b - \Delta$, and the zone of preference for acceptance of H_3 by $\theta \geqq b + \Delta$. The zone of indifference between H_1 and H_2 is given by the inequality $a - \Delta < \theta < a + \Delta$, the zone of indifference between H_1 and H_3 is empty, and the zone of indifference between H_2 and H_3 is given by $b - \Delta \leqq \theta < b + \Delta$. Finally, the zone of indifference among H_1, H_2, and H_3 is empty.

When the weight functions $\bar{w}_1(\theta), \cdots, \bar{w}_k(\theta)$ are used, the risk function $r(\theta)$ defined in (10:2) takes a particularly simple form. Since $\bar{w}_j(\theta)$ can take only the values 0 and 1, we shall have

$$(10{:}5) \qquad\qquad r(\theta) = \sum_j L_j(\theta)$$

where the summation is to be taken for all values of j for which $\bar{w}_j(\theta) = 1$.

We shall say that a wrong decision is made if, and only if, a hypothesis H_i is accepted for which $\bar{w}_i(\theta) = 1$. Then the risk $r(\theta)$ given in (10:5) is simply equal to the probability that a wrong decision will be made.

The principle for the selection of a sequential sampling plan, as stated in Section 10.4.3, can now be formulated as follows. We consider only sequential sampling plans for which the probability of making a wrong decision does not exceed a certain preassigned value r_0. From the class of such sequential sampling plans we try to select one for which the expected value of the number of observations required by the plan is as small as possible.

10.5 Discussion of a Special Class of Sequential Sampling Plans

The problem of finding a sequential sampling plan which may be regarded as an optimum plan in the sense of the previous section is not yet solved. However, as will be shown in this section, a wide class of sequential sampling plans can be constructed for which the condition that the probability of making a wrong decision should not exceed a preassigned value r_0 is fulfilled.

To construct such a class of sampling plans we shall make use of the following lemma.

Lemma. Let $x_1, x_2, \cdots$, etc., be a sequence of variates, let $p_{1m}(x_1, \cdots, x_m)$ $(m = 1, 2, \cdots)$ denote the joint probability density function of $x_1, \cdots, x_m$ under the hypothesis H_1, and let $p_{0m}(x_1, \cdots, x_m)$ be the density function under the hypothesis H_0.[4] Let, furthermore, A be a constant greater than one. Then, under the hypothesis H_0, the probability that

$$(10:6) \qquad \frac{p_{1m}(x_1, \cdots, x_m)}{p_{0m}(x_1, \cdots, x_m)} < A$$

will hold for all values of m is greater than or equal to $1 - (1/A)$.

The validity of this lemma can easily be shown with the help of the inequalities given in Section 3.2 by letting the constant B in those inequalities approach 0.

With the help of this lemma we can construct a sequential sampling plan satisfying the condition that the probability of making a wrong decision does not exceed a prescribed value r_0 as follows. Let $p_m(x_1, \cdots, x_m, \theta)$ be equal to $f(x_1, \theta)f(x_2, \theta) \cdots f(x_m, \theta)$ where $f(x, \theta)$ is the probability distribution of x when θ is true. For any parameter point θ let $p_m{}^*(x_1, \cdots, x_m, \theta)$ be an arbitrary but given probability distribution of the variates $x_1, x_2, \cdots, x_m$.[5] Then according to our lemma the probability that

$$(10:7) \qquad \frac{p_m{}^*(x_1, \cdots, x_m, \theta)}{p_m(x_1, \cdots, x_m, \theta)} < A$$

will hold for all m is greater than or equal to $1 - (1/A)$ when θ is true. For any sample point $E_n = (x_1, \cdots, x_n)$, let $\omega_n(E_n)$ denote the totality of all parameter points θ for which the inequality (10:7) is fulfilled for all values $m \leq n$. Clearly, the probability that the true parameter point θ will be included in all sets $\omega_n(E_n)$ $(n = 1, 2, \cdots$, ad inf.) is greater than or equal to $1 - (1/A)$. The sequential sampling plan is then defined as follows: We continue taking additional observations as long as none of the weight functions $\bar{w}_1(\theta), \cdots, \bar{w}_k(\theta)$ is identically zero in $\omega_n(E_n)$. At the first time when $\omega_n(E_n)$ is such that at least one

---

[4] If the distribution of $x_1, x_2, \cdots$, etc. is discrete, $p_{im}(x_1, \cdots, x_m)$ denotes the probability of obtaining a sample equal to the observed.

[5] It is understood that the distribution of $x_1, \cdots, x_m$ determined from the distribution $p_{m'}{}^*(x_1, \cdots, x_{m'}, \theta)$ $(m' > m)$ is identical with $p_m{}^*(x_1, \cdots, x_m, \theta)$.

of the weight functions $\bar{w}_1(\theta), \cdots, \bar{w}_k(\theta)$ is identically 0 in $\omega_n(E_n)$, we stop the process with the acceptance of the hypothesis corresponding to the weight function which is identically zero in $\omega_n(E_n)$.[6] Obviously, this sequential sampling plan will have the property that the probability of making a wrong decision does not exceed $1/A$. If we let A equal $1/r_0$, then the probability of making a wrong decision will not exceed r_0, as required.

This method leads to a wide class C of sequential sampling plans with the required property, since the distribution function $p_m^*(x_1, \cdots, x_m, \theta)$ in the numerator of (10:7) can be chosen entirely arbitrarily. It is doubtful whether this class C of sampling plans contains an optimum plan in the sense of the definition given in 10.4. If we are willing to restrict ourselves to sampling plans in class C, we still have the problem of so choosing $p_m^*(x_1, \cdots, x_m, \theta)$ as to make the expected number of observations required by the plan as small as possible. This problem, too, has not yet been solved. There may be some waste involved in letting $A = 1/r_0$, since this may result in a maximum probability of making a wrong decision that is considerably less than the tolerated value r_0. A further development of the theory may show that A can be put equal to some value smaller than $1/r_0$ which would lead to a saving in the number of observations.

Although the present stage of the theory is very incomplete, sampling plans based on the inequality (10:7) may still be used with good advantage in some problems. Even if we cannot yet find the best distribution $p_m^*(x_1, \cdots, x_m, \theta)$ to be used in the numerator of (10:7), we still may be able to make a reasonably good choice of $p_m^*(x_1, \cdots, x_m, \theta)$ and thereby obtain a sequential plan which requires, on the average, a substantially smaller number of observations than the best possible non-sequential sampling plan based on a predetermined number of observations.

Regarding possible choices of $p_m^*(x_1, \cdots, x_m, \theta)$ which may give reasonably good results, the following remarks may be made. A good result may be obtained in some problems by letting $p_m^*(x_1, \cdots, x_m, \theta)$ equal a properly chosen weighted average of $p_m(x_1, \cdots, x_m, \zeta)$ where ζ is a variable parameter point. In other words, we let [7]

$$(10:8) \qquad p_m^*(x_1, \cdots, x_m, \theta) = \int_\Omega \rho_\theta(\zeta) p_m(x_1, \cdots, x_m, \zeta)\, d\zeta$$

[6] If there are several weight functions which are identically 0 in $\omega_n(E_n)$, we may choose arbitrarily one from among the hypotheses corresponding to these weight functions.

[7] The averaging function $\rho_\theta(\zeta)$ may also be discrete. Formulas valid for both continuous and discrete averaging functions could be given by using Stieltje's integrals.

where the integration is taken over the whole parameter space Ω and $\rho_\theta(\zeta)$ is a non-negative function of ζ satisfying the condition

$$(10:9) \qquad \int_\Omega \rho_\theta(\zeta)\, d\zeta = 1$$

The choice of the averaging function $\rho_\theta(\zeta)$ will depend on the weight functions $\bar{w}_1(\theta), \cdots, \bar{w}_k(\theta)$. If, for example, $\bar{w}_j(\theta) = 0$ for the parameter point θ under consideration, it seems reasonable to let $\rho_\theta(\zeta) = 0$ for all parameter points ζ for which $\bar{w}_j(\zeta) = 0$, since we are not interested in discriminating between parameter points for which the same decision is correct.

The following is another possible choice of $p_m*(x_1, \cdots, x_m, \theta)$ which may lead to good results in some problems:

$$(10:10) \quad p_m*(x_1, \cdots, x_m, \theta) = \phi(x_1, \theta)f(x_2, \hat{\theta}_1)f(x_3, \hat{\theta}_2) \cdots f(x_m, \hat{\theta}_{m-1})$$

where $\hat{\theta}_r$ is the maximum likelihood estimate of θ based on the first r observations $x_1, \cdots, x_r$ and $\phi(x_1, \theta)$ is some suitably chosen probability distribution of x_1.

To illustrate the sampling procedure based on (10:7), we shall consider the following simple example. Let x be normally distributed with unknown mean θ and unit variance. Then

$$(10:11) \qquad p_m(x_1, \cdots, x_m, \theta) = \frac{1}{(2\pi)^{\frac{m}{2}}} e^{-\frac{1}{2} \sum_{\alpha=1}^m (x_\alpha - \theta)^2}$$

Let

$$(10:12) \quad p_1*(x_1, \cdots, x_m, \theta)$$
$$= \tfrac{1}{2}[p_m(x_1, \cdots, x_m, \theta + \delta) + p_m(x_1, \cdots, x_m, \theta - \delta)]$$

where δ is a given positive quantity. Then

$$(10:13) \quad \frac{p_m*(x_1, \cdots, x_m, \theta)}{p_m(x_1, \cdots, x_m, \theta)} = \frac{e^{-\frac{1}{2}m\delta^2}}{2} [e^{\delta\Sigma(x_\alpha-\theta)} + e^{-\delta\Sigma(x_\alpha-\theta)}]$$
$$= e^{-\frac{1}{2}m\delta^2} \cosh [\delta\Sigma(x_\alpha - \theta)]$$

The equation

$$(10:14) \qquad \cosh u = v \qquad (v > 1)$$

has two roots in u which are equal in absolute value. Let $\psi(v)$ be the positive, and $-\psi(v)$ the negative root of (10:14). Then the roots of the equation in θ

$$(10:15) \qquad e^{-\frac{1}{2}m\delta^2} \cosh [\delta\Sigma(x_\alpha - \theta)] = A$$

are given by

(10:16)
$$
\begin{cases}
\theta_1(E_m) = \bar{x}_m + \dfrac{\psi(e^{\frac{m\delta^2}{2}} A)}{m\delta} \\[2ex]
\text{and} \\[2ex]
\theta_2(E_m) = \bar{x}_m - \dfrac{\psi(e^{\frac{m\delta^2}{2}} A)}{m\delta}
\end{cases}
$$

where $\bar{x}_m$ is the arithmetic mean of the observations $x_1, \cdots, x_m$.

The set of all values of θ for which the inequality

$$
\frac{p_m^*(x_1, \cdots, x_m, \theta)}{p_m(x_1, \cdots, x_m, \theta)} < A
$$

is satisfied is the open interval $(\theta_2(E_m), \theta_1(E_m))$. The set $\omega_n(E_n)$ is defined as the common part of the open intervals $(\theta_2(E_1), \theta_1(E_1)), \cdots,$ $(\theta_2(E_n), \theta_1(E_n))$. Hence $\omega_n(E_n)$ is equal to the open interval whose lower endpoint is equal to the maximum of the values $\theta_2(E_1), \cdots,$ $\theta_2(E_n)$, and whose upper endpoint is equal to the minimum of the values $\theta_1(E_1), \cdots, \theta_1(E_n)$.[8] Experimentation is terminated the first time the open interval $\omega_n(E_n)$ is such that one of the weight functions $\bar{w}_1(\theta), \cdots, \bar{w}_k(\theta)$ is identically zero in $\omega_n(E_n)$.

As another illustration, consider again the example given in Section 10.1, and for simplicity assume that the standard deviation of x is equal to 1. Although the proper choice of $p_m^*(x_1, \cdots, x_m, \theta)$ for this example has not been thoroughly investigated, the following choice of $p_m^*(x_1, \cdots, x_m, \theta)$ is perhaps not unreasonable. A parameter point θ in the zone of preference for acceptance of H_1, i.e., a value $\theta \leq a - \Delta$,[9] should be discriminated against all other parameter values ζ for which acceptance of H_1 is a wrong decision. The smallest value ζ for which acceptance of H_1 is a wrong decision, i.e., the smallest ζ for which $\bar{w}_1(\zeta) = 1$, is $\zeta = a + \Delta$. Thus, we put

(10:17) $p_m^*(x_1, \cdots, x_m, \theta) = p_m(x_1, \cdots, x_m, a + \Delta)$

for all $\theta \leq a - \Delta$

If θ is in the zone of indifference between H_1 and H_2, i.e., if $a - \Delta < \theta < a + \Delta$, we want to discriminate θ against values ζ for which ac-

---

[8] If it happens that the upper endpoint determined in this way is less than the lower endpoint, the set $\omega_n(E_n)$ is empty.

[9] For a definition of the various zones and weight functions $\bar{w}_1(\theta)$, $\bar{w}_2(\theta)$, and $\bar{w}_3(\theta)$ for this example see Section 10.4.4.

ceptance of H_1, as well as of H_2, is a wrong decision. The smallest value of this kind is $\zeta = b + \Delta$. Thus, we let

$$(10{:}18) \quad p_m^*(x_1, \cdots, x_m, \theta) = p_m(x_1, \cdots, x_m, b + \Delta)$$

$$\text{if } a - \Delta < \theta < a + \Delta$$

If θ is in the zone of preference for acceptance of H_2, i.e., if $a + \Delta \leqq \theta < b - \Delta$, we want to discriminate it against values ζ for which acceptance of H_2 is wrong. The greatest value ζ of this kind to the left of $a + \Delta$ is $\zeta = a - \Delta$, and the smallest ζ of this kind to the right of $b - \Delta$ is $\zeta = b + \Delta$. It seems, therefore, reasonable to let

$$(10{:}19) \quad p_m^* = \tfrac{1}{2}[p_m(x_1, \cdots, x_m, a - \Delta) + p_m(x_1, \cdots, x_m, b + \Delta)]$$

$$\text{if } a + \Delta \leqq \theta < b - \Delta$$

If θ is in the zone of indifference between H_2 and H_3, i.e., if $b - \Delta \leqq \theta < b + \Delta$, we want to discriminate θ against values ζ for which the acceptance of H_2, as well as of H_3, is wrong. Thus, we let

$$(10{:}20) \quad p_m^*(x_1, \cdots, x_m, \theta) = p_m(x_1, \cdots, x_m, a - \Delta)$$

$$\text{if } b - \Delta \leqq \theta < b + \Delta$$

Finally, if θ is in the zone of preference for acceptance of H_3, i.e., if $\theta \geqq b + \Delta$, we want to discriminate θ against values ζ for which the acceptance of H_3 is wrong. The least upper bound of values of ζ of this kind is $\zeta = b - \Delta$. Thus, we shall let

$$(10{:}21) \quad p_m^*(x_1, \cdots, x_m, \theta) = p_m(x_1, \cdots, x_m, b - \Delta)$$

$$\text{for } \theta \geqq b + \Delta$$

It should be remembered that there is no systematic theory yet available for the proper choice of $p_m^*(x_1, \cdots, x_m, \theta)$. The choice of $p_m^*(x_1, \cdots, x_m, \theta)$ in the above example has been made only on intuitive grounds. It may well be that another choice of $p_m^*(x_1, \cdots, x_m, \theta)$ exists which leads to much better results. It should also be remarked that it is doubtful whether an optimum sampling plan, as defined in the preceding section, is a member of the class of sampling plans based on the inequality (10:7). Further investigations are needed to clarify these questions.

Chapter 11. THE PROBLEM OF SEQUENTIAL ESTIMATION

11.1 Principles of the Current Theory of Estimation by Intervals or Sets

In this section we shall give a brief outline of the basic ideas of estimation by intervals or sets as developed by J. Neyman.[1] Consider first the case in which the distribution of the random variable x under consideration is known except for the value of a single parameter θ. The problem treated in the current theory is that of estimating the value of θ on the basis of a fixed number of observations, say N observations $x_1, \cdots, x_N$ on x.

Let E denote the sample $(x_1, \cdots, x_N)$ and let $\underline{\theta}(E)$ and $\bar{\theta}(E)$ be two single-valued functions of the sample E such that

$$(11:1) \qquad \underline{\theta}(E) \leqq \bar{\theta}(E) \qquad \text{for all possible samples } E$$

Let $\delta(E)$ denote the interval extending from $\underline{\theta}(E)$ to $\bar{\theta}(E)$. We shall refer to $\delta(E)$ also as an interval function, since it associates an interval with each sample. Since the interval $\delta(E)$ is a function of the sample, its location and length will, in general, be random variables and, therefore, probability statements can be made as to whether $\delta(E)$ includes the true parameter value θ or not. For any value θ we shall express the relation that $\delta(E)$ contains θ by the symbol $\delta(E)C\theta$. For any relation R, the symbol $P(R \mid \theta)$ will denote the probability that R holds when θ is the true parameter value.

According to Neyman, an interval function $\delta(E)$ is said to be a confidence interval of θ if

$$(11:2) \qquad P[\delta(E)C\theta \mid \theta] = \gamma$$

identically in θ where γ is a fixed value independent of θ. The relation (11:2) simply says this: The probability that $\delta(E)$ will include the true parameter value is always equal to γ no matter what the true value of the parameter happens to be. The fixed value γ is called the confidence coefficient associated with the confidence interval $\delta(E)$.

---

[1] J. Neyman, "Outline of a Theory of Statistical Estimation Based on the Classical Theory of Probability," *Philosophical Transactions of the Royal Society of London*, Series A, Vol. 236 (1937), pp. 333–380.

Suppose, now, that the distribution of x involves several unknown parameters, say $\theta_1, \cdots, \theta_r$. Any set of possible values $\theta_1, \cdots, \theta_r$ can be represented by a point θ, called a parameter point, in the r-dimensional Cartesian space (parameter space). If we want to estimate the parameters $\theta_1, \cdots, \theta_r$ jointly, i.e., if we want to estimate the parameter point θ, the estimating set will be some subset of the r-dimensional parameter space. Whereas in the case of a single unknown parameter, estimating sets other than intervals have little practical value, this is not so when several unknown parameters are to be estimated jointly. Estimating sets other than intervals in the r-dimensional space, such as the interior of a sphere, or ellipse, or more general regions, will have to be considered. Thus, we shall have to consider a set function $\omega(E)$ which associates with each sample point E a certain subset $\omega(E)$ of the parameter space without making the restriction that $\omega(E)$ is an r-dimensional interval.

A set function $\omega(E)$ is said to be a confidence region of the parameter point $\theta = (\theta_1, \cdots, \theta_r)$ if

$$(11\!:\!3) \qquad\qquad P[\omega(E)C\theta \mid \theta] = \gamma$$

identically in θ where γ is a fixed value independent of θ. The value γ is called the confidence coefficient of the confidence region $\omega(E)$.

If only one of the parameters $\theta_1, \cdots, \theta_r$ is to be estimated, estimating sets other than one-dimensional intervals will not be of much practical interest, as in the case of a single unknown parameter. Suppose, for example, that only θ_1 is to be estimated. According to Neyman, an interval function $\delta(E)$ is said to be a confidence interval of θ_1 with confidence coefficient γ if

$$(11\!:\!4) \qquad\qquad P[\delta(E)C\theta_1 \mid \theta_1, \theta_2, \cdots, \theta_r] = \gamma$$

identically in $\theta_1, \theta_2, \cdots, \theta_r$.

Usually there will be infinitely many confidence intervals $\delta(E)$ or confidence regions $\omega(E)$ with a given confidence coefficient γ and a fundamental problem is to find a proper confidence interval or confidence region which has some optimum properties. It is clear that a confidence interval or confidence region with a given confidence coefficient γ will be regarded the better the shorter the interval or the smaller the region. The notion "short" or "small" is to be made precise, since the length of a confidence interval and the size of a confidence region are random variables depending on the outcome of the sample. This has been done in the theory developed by Neyman who introduced various notions of optimum confidence intervals and confidence regions. The mathematical consequences of these definitions have been investigated and optimum confidence intervals and regions have been derived in many important cases. It is not intended to go into further details here and the reader is referred to the original publications of Neyman on this subject.

11.2 Formulation of the Problem of Sequential Estimation by Intervals or Sets

In estimation procedures based on a fixed number of observations, we cannot control, in general, the length of the confidence interval obtained, since this depends on the outcome of the sample. It may, therefore, sometimes happen that the confidence interval obtained is so long that it has little or no practical value. The possibility of such an event is a drawback inherent in estimation procedures based on a predetermined number of observations.

For example, the length of the best confidence interval, based on a fixed number of observations, for the mean of a normal population with unknown standard deviation is proportional to the sample estimate s of the population standard deviation σ. The sample standard deviation s may take any value and is likely to be large if σ is large.

To devise estimation procedures which lead to confidence intervals not only with a prescribed confidence coefficient but also with a prescribed length, or with a length not exceeding a prescribed value, or which satisfies some other similar condition, it is, in general, necessary to abandon the approach based on a fixed number of observations, and estimation procedures of sequential nature have to be constructed.[2]

The general nature of a sequential procedure of estimation by sets may be described as follows. For any positive integer m we consider a set S_m of samples of size m. These sets must satisfy the following condition. If the sample E_m is an element of S_m and if $E_{m'}$ $(m' > m)$ is an element of $S_{m'}$, then E_m must not be equal to the sample consisting of the first m observations in $E_{m'}$. With any element E_m of S_m $(m = 1, 2, \cdots, \text{ad inf.})$, we associate a subset $\omega(E_m)$ of the parameter space.[3] The sequential process of estimation is then carried out as follows. We continue to make observations on x until we reach a value n such that E_n is an element of S_n. At this stage, we stop the process

---

[2] A very interesting sequential procedure has been devised by C. Stein, "A Two Sample Test for a Linear Hypothesis whose Power Is Independent of the Variance," *The Annals of Mathematical Statistics*, Vol. XVI, Sept., 1945, which leads to confidence intervals of fixed length in an important class of problems, including the example mentioned before.

[3] If we are concerned with interval estimation, $\omega(E_m)$ will always be an interval.

and state that $\omega(E_n)$ contains the true parameter point, i.e., $\omega(E_n)$ is the confidence set resulting from the sequential estimation procedure.

Thus, a sequential estimation procedure is determined by the sample sets S_1, S_2, $\cdots$, etc., and the set function $\omega(E)$ defined for all samples E in S_1, S_2, $\cdots$, etc. The fundamental problem in sequential estimation is that of a proper choice of S_1, S_2, $\cdots$, etc., and of $\omega(E)$. First we impose the following two conditions:

Condition I. The confidence set $\omega(E_n)$ resulting from the sequential estimation procedure should satisfy certain stated requirements regarding its geometric shape.

Condition II. The confidence set $\omega(E_n)$ resulting from the sequential estimation procedure should satisfy the inequality [4]

$$P[\omega(E_n)C\theta \mid \theta] \geqq \gamma$$

for all parameter points θ. (The quantity γ is a fixed value which is frequently chosen as high as .95, or more.)

The requirements to be imposed on the geometric shape of the confidence set $\omega(E_n)$ do not constitute a statistical problem, and they will be decided on the basis of practical considerations in each particular problem. For example, if there is only one unknown parameter θ (the parameter space is one-dimensional), we may want to require that $\omega(E)$ be an interval whose length should not exceed some fixed prescribed value d, or some given function of the midpoint of the interval. The latter case may be of interest, for example, in estimating the mean of a binomial distribution. If there are several unknown parameters, say θ_1, $\cdots$, θ_r, and we want to estimate them jointly, we may require that the Euclidean volume, or the diameter [5] of the confidence set $\omega(E_n)$ does not exceed some fixed prescribed value. If we merely want to estimate one of the unknown parameters, say θ_1, we may impose the requirement that $\omega(E_n)$ be an interval with length not exceeding some prescribed fixed value, or the weaker requirement that $\omega(E_n)$ be a subset of the r-dimensional parameter space whose projection on the θ_1-axis has a diameter not exceeding some preassigned value.

Usually there will exist infinitely many sequential estimation procedures which satisfy Conditions I and II. The criterion for selecting one from among them will be based on the expected number of obser-

---

[4] This is weaker than the requirement by Neyman that the equality sign should hold.

[5] The diameter of a set is the largest possible distance between two points of the set.

vations required by the estimation procedure. The sequential estimation procedure may be regarded the better the smaller the expected
number of observations required by the procedure. Thus, we shall try
to select a sequential estimation procedure from the class of procedures
satisfying Conditions I and II for which the expected number of observations to be made is as small as possible.

The problem of finding an optimum estimation procedure is un-
solved. However, a special class of estimation procedures satisfying
Conditions I and II will be discussed briefly in the next section. It is
doubtful whether this class of procedures contains an optimum solution in the sense defined before.

11.3 A Special Class of Sequential Estimation Procedures

The special class of sampling plans based on the inequality (10:7),
and discussed in Section 10.5, can be used to obtain estimation procedures satisfying Conditions I and II. With each sample point $E_n =
(x_1, \cdots, x_n)$ $(n = 1, 2, \cdots,$ ad inf.) we associate the set $\omega(E_n)$ con-
sisting of all parameter points θ for which (10:7) is fulfilled for all
values $m \leqq n$. If we put $A = 1/(1 - \gamma)$, then $\omega(E_n)$ will satisfy Condition II for each n. The estimation procedure is carried out as follows. We continue taking observations as long as $\omega(E_n)$ does not
satisfy the requirements in Condition I. We stop the process at the
smallest n for which $\omega(E_n)$ satisfies Condition I and then state that
the true parameter point θ is included in $\omega(E_n)$. This rule of stopping
insures automatically the fulfillment of Condition I.

If $p_m^*(x_1, \cdots, x_m, \theta)$ is chosen so that the probability is 1 that the
diameter of $\omega(E_m)$ will converge to 0 as $m \to \infty$, and if Condition I is
such that any set of sufficiently small diameter satisfies it, the probability is 1 that the estimation process will be terminated at a finite
stage.

It is doubtful whether the special class of procedures considered here
contains an optimum procedure in the sense of the preceding section.
Even if we are willing to restrict ourselves to procedures based on
(10:7), there is no theory yet developed for the proper choice of
$p_m^*(x_1, \cdots, x_m, \theta)$. Our aim is, of course, to choose $p_m^*(x_1, \cdots, x_m, \theta)$
so that the expected number of observations required by the pro-
cedure should be as small as possible. An optimum choice of
$p_m^*(x_1, \cdots, x_m, \theta)$ will depend also on the nature of Condition I.
For example, if a certain choice of $p_m^*(x_1, \cdots, x_m, \theta)$ is optimal when
Condition I requires that the diameter of $\omega(E_n)$ does not exceed a preassigned value, this choice will probably not be optimal when Condition

I requires that the diameter of the projection of $\omega(E_n)$ on one of the parameter axes does not exceed a preassigned value, and vice versa.

There may be some waste involved in putting $A = 1/(1 - \gamma)$, since this may imply the validity of Condition II for a value γ' substantially larger than the intended γ. A further development of the theory may show that A can be put equal to some value smaller than $1/(1 - \gamma)$ which would lead to a saving in the number of observations.

APPENDIX

A.1 PROOF THAT THE PROBABILITY IS 1 THAT THE SEQUENTIAL PROBABILITY RATIO TEST WILL EVENTUALLY TERMINATE

The sequential probability ratio test terminates at the nth trial where n is the smallest integer for which either

$$z_1 + \cdots + z_n \geqq \log A$$

or

$$z_1 + \cdots + z_n \leqq \log B \qquad \left[z_i = \log \frac{f(x_i, \theta_1)}{f(x_i, \theta_0)} \right]$$

Let $c = |\log B| + |\log A|$. We shall subdivide the infinite sequence $z_1, z_2, z_3, \cdots$, ad inf., into segments of length r where r is some positive integer. Thus, the first segment S_1 will consist of the elements $z_1, \cdots, z_r$, the second segment S_2 will contain the elements $z_{r+1}, \cdots, z_{2r}$, etc. In general, the kth segment S_k will consist of the elements $z_{(k-1)r+1}, \cdots, z_{kr}$. Let ζ_k denote the sum of the elements in the kth segment. It can be seen that if the infinite sequence $z_1, z_2, \cdots$, ad inf., is such that the sequential process never terminates, then we must have

(A:1) $$|\zeta_k| < c \qquad \text{for } k = 1, 2, \cdots, \text{ad inf.}$$

Inequality (A:1) can also be written

(A:2) $$(\zeta_k)^2 < c^2 \qquad \text{for } k = 1, \cdots, \text{ad inf.}$$

Thus, in order to show that the probability is 1 that the sequential process will eventually terminate, it is sufficient to prove that the probability is 0 that (A:2) holds for all integral values k. For any given positive integer i denote by P_i the probability that $\zeta_i^2 < c^2$. Since $z_1, z_2, \cdots$, are independently distributed, each having the same distribution, the distribution of ζ_i must be the same for all values i. Hence, also P_i is independent of i and we shall denote it by P. Since $\zeta_1, \zeta_2, \cdots$, etc., are independently distributed, the probability of the joint event that (A:2) holds for $k = 1, 2, \cdots, j$ is equal to P^j. Hence, in order to show that the probability is 0 that (A:2) holds for *all* values k, it is sufficient to show that $P < 1$. Clearly, if the expected value of ζ_i^2 is $> c^2$, then P must be < 1. Since the variance of z_i is assumed to be positive, the expected value of ζ_i^2 can be made arbitrarily large by choosing r, i.e., the number of elements in a segment, sufficiently large. Thus, $P < 1$, and we have proved the proposition: *The probability is 1 that the sequential probability ratio test procedure will eventually terminate.*

A.2   UPPER AND LOWER LIMITS FOR THE OC FUNCTION OF A SEQUENTIAL TEST
A.2.1   A Lemma

In what follows we shall denote the expected value of any random variable z by $E(z)$. For any relation R we shall use the symbol $P(R)$ to denote the probability that R holds. If the expected value $E(z)$ or the probability $P(R)$ has been determined under the assumption that θ is the true value of the parameter involved in the distribution of the random variable under consideration, we shall occasionally put this in evidence by using the symbols $E_\theta(z)$ and $P_\theta(R)$, respectively.[1]

In deriving lower and upper limits for the OC function of a sequential test, we shall make use of the following lemma.

Lemma A.1. Let z be a random variable such that the following three conditions are fulfilled:

Condition I.   The expected value $E(z)$ exists and is not equal to 0.

Condition II.   There exists a positive δ such that $P(e^z < 1 - \delta) > 0$ and $P(e^z > 1 + \delta) > 0$.

Condition III.   For any real value h the expected value $E(e^{hz}) = g(h)$ exists.

Then there exists one and only one real value $h_0 \neq 0$ such that

$$E(e^{h_0 z}) = 1$$

Proof: For any positive h we have

(A:3) $$g(h) > P(e^z > 1 + \delta)(1 + \delta)^h$$

Hence, since $P(e^z > 1 + \delta) > 0$,

(A:4) $$\lim_{h=\infty} g(h) = +\infty$$

Similarly, we see that for any negative h

$$g(h) > P(e^z < 1 - \delta)(1 - \delta)^h$$

Hence, since $P(e^z < 1 - \delta) > 0$, we have

(A:5) $$\lim_{h=-\infty} g(h) = +\infty$$

[1] If there are several unknown parameters, say $\theta_1, \cdots, \theta_k$, then θ denotes the set $(\theta_1, \cdots, \theta_k)$.

Since $g''(h) = E(z^2 e^{hz})$,[2] it follows from Condition II that

(A:6) $$g''(h) > 0$$

for all real values of h.

The relations (A:4), (A:5), (A:6) imply that there exists exactly one real value h^* such that $g(h)$ takes its minimum value for $h = h^*$. Since $g'(0) = E(z)$ is unequal to 0 by Condition I, we see that $h^* \neq 0$ and $g(h^*) < g(0) = 1$. It is clear that the function $g(h)$ is monotonically decreasing in the strict sense over the interval $(-\infty, h^*)$ and is monotonically increasing in the strict sense over the interval $(h^*, +\infty)$. Since $g(0) = 1$ and $g(h^*) < 1$, there exists exactly one real value $h_0 \neq 0$ such that $g(h_0) = 1$. Hence lemma A.1 is proved.

From the above considerations it follows that if $h^* > 0$ then also $h_0 > 0$, and if $h^* < 0$ then also $h_0 < 0$. Furthermore, if $h^* > 0$ then $E(z) = g'(0) < 0$, and if $h^* < 0$ then $E(z) = g'(0) > 0$. Hence, h_0 and $E(z)$ are of opposite sign.

A.2.2 A Fundamental Identity

In this section we shall derive an identity which will play a fundamental role. Consider the sequential probability ratio test for testing the hypothesis H_0 that the probability distribution of x is given by $f(x, \theta_0)$ against the alternative hypothesis H_1 that the probability distribution in question is given by $f(x, \theta_1)$. Let $z = \log \dfrac{f(x, \theta_1)}{f(x, \theta_0)}$ and $z_i = \log \dfrac{f(x_i, \theta_1)}{f(x_i, \theta_0)}$ where x_i denotes the ith observation on x. As defined in Section 3.1, the test procedure is given as follows. Continue taking observations as long as

(A:7) $$\log B < z_1 + \cdots + z_m < \log A$$

where A and B $(B < A)$ are constants determined before the experimentation starts. Accept H_0 when

(A:8) $$z_1 + \cdots + z_m \leqq \log B$$

and reject H_0 (accept H_1) when

(A:9) $$z_1 + \cdots + z_m \geqq \log A$$

---

[2] From Condition III it follows that all derivatives of $g(h)$ exist, and they may be obtained by differentiation under the integral sign, i.e.,

$$\frac{d^r g(h)}{dh^r} = E(z^r e^{zh}) \qquad (r = 1, 2, \cdots, \text{ad inf.})$$

In what follows we shall denote by n the number of observations required by the test. Clearly, n is a random variable. Let D' be the subset of the complex plane such that $E(e^{zt}) = \phi(t)$ exists and is finite for any point t in D'. Consider the following identity:

$$(A{:}10) \qquad E(e^{Z_n t + (Z_N - Z_n)t}) = E(e^{Z_N t}) = [\phi(t)]^N$$

where N denotes a positive integer and $Z_i = z_1 + \cdots + z_i$. Let P_N be the probability that $n \leqq N$. For any random variable u, let $E_N(u)$ denote the conditional expected value of u under the restriction that $n \leqq N$, and let $E_N{}^*(u)$ denote the conditional expected value of u under the restriction that $n > N$. Then identity (A:10) can be written as

$$(A{:}11) \quad P_N E_N(e^{Z_n t + (Z_N - Z_n)t}) + (1 - P_N)E_N{}^*(e^{Z_N t}) = [\phi(t)]^N$$

Since in the subpopulation defined by any fixed $n \leqq N$ the expression $Z_N - Z_n$ is independent of Z_n, we have

$$(A{:}12) \qquad E_N(e^{Z_n t + (Z_N - Z_n)t}) = E_N\{(e^{Z_n t})[\phi(t)]^{N-n}\}$$

From (A:11) and (A:12) we obtain the identity

$$(A{:}13) \quad P_N E_N\{e^{Z_n t}[\phi(t)]^{N-n}\} + (1 - P_N)E_N{}^*(e^{Z_N t}) = [\phi(t)]^N$$

Dividing both sides by $[\phi(t)]^N$ we obtain

$$(A{:}14) \qquad P_N E_N\{e^{Z_n t}[\phi(t)]^{-n}\} + (1 - P_N) \frac{E_N{}^*(e^{Z_N t})}{[\phi(t)]^N} = 1$$

Let D'' be the subset of the complex plane in which $\big| \phi(t) \big| \geqq 1$ and let D denote the common part of the subsets D' and D''. Since $\lim_{N = \infty} (1 - P_N) = 0$, and since $\big| E_N{}^*(e^{Z_N t}) \big|$ is a bounded function of N, we have in D

$$(A{:}15) \qquad \lim_{N = \infty} (1 - P_N) \frac{E_N{}^*(e^{Z_N t})}{[\phi(t)]^N} = 0$$

Since

$$\lim_{N = \infty} P_N E_N\{e^{Z_n t}[\phi(t)]^{-n}\} = E\{e^{Z_n t}[\phi(t)]^{-n}\}$$

we obtain from (A:14) and (A:15) the fundamental identity

$$(A{:}16) \qquad E\{e^{Z_n t}[\phi(t)]^{-n}\} = 1$$

for any point t in the set D.

A.2.3   Derivation of Upper and Lower Limits for the OC Function

The OC function of the sequential test is defined by the function $L(\theta)$, where $L(\theta)$ denotes the probability that the sequential process leads to the acceptance of H_0 when θ is the true value of the parameter.[3] It has been shown in Section A.1 that the probability is 0 that the sequential process will never terminate, i.e., the relation $P(n = \infty) = 0$ has been proved. Thus, the probability that the process will terminate with the rejection of H_0 (acceptance of H_1) is given by $1 - L(\theta)$. Using the fundamental identity derived in the preceding section we shall obtain upper and lower limits for $L(\theta)$.

It will be assumed that the distribution of $z = \log \dfrac{f(x, \theta_1)}{f(x, \theta_0)}$ satisfies the three conditions of lemma A.1 for any value θ. Then for any given θ there exists exactly one real value $h(\theta) \neq 0$ such that $E_\theta(e^{zh(\theta)}) = 1$. Substituting $h(\theta)$ for t in the fundamental identity (A:16), we obtain

$$(\text{A:17}) \qquad\qquad E_\theta(e^{Z_n h(\theta)}) = 1$$

since $\phi[h(\theta)] = 1$.

Let $E_\theta{}^*$ be the conditional expected value of $e^{Z_n h(\theta)}$ under the restriction that H_0 is accepted, i.e., that $Z_n \leq \log B$, and let $E_\theta{}^{**}$ be the conditional expected value of $e^{Z_n h(\theta)}$ under the restriction that H_1 is accepted, i.e., that $Z_n \geq \log A$. Then we obtain, from (A:17),

$$(\text{A:18}) \qquad\qquad [L(\theta)]E_\theta{}^* + [1 - L(\theta)]E_\theta{}^{**} = 1$$

Solving for $L(\theta)$ we obtain

$$(\text{A:19}) \qquad\qquad L(\theta) = \frac{E_\theta{}^{**} - 1}{E_\theta{}^{**} - E_\theta{}^*}$$

If both the absolute value of $E_\theta(z)$ and the variance of z are small, as they will be when $f(x, \theta_1)$ is near $f(x, \theta_0)$, then $E_\theta{}^*$ and $E_\theta{}^{**}$ will be nearly equal to $B^{h(\theta)}$ and $A^{h(\theta)}$, respectively. Hence, in this case a good approximation to $L(\theta)$ is given by the expression

$$(\text{A:20}) \qquad\qquad \bar{L}(\theta) = \frac{A^{h(\theta)} - 1}{A^{h(\theta)} - B^{h(\theta)}}$$

This is the approximation formula (3:43) given in Section 3.4. It is easy to verify that $h(\theta) = 1$ if $\theta = \theta_0$, and $h(\theta) = -1$ if $\theta = \theta_1$. The difference $L(\theta) - \bar{L}(\theta)$ approaches 0 if both the mean and the variance of z converge to 0.

---

[3] For simplicity the case of a single unknown parameter θ is discussed, but the results can obviously be extended to any number of parameters.

To judge the goodness of the approximation given by $\bar{L}(\theta)$, it is desirable to derive lower and upper limits for $L(\theta)$. Such limits can be obtained by deriving lower and upper limits for $E_\theta{}^*$ and $E_\theta{}^{**}$. First we consider the case when $h(\theta) > 0$. To obtain a lower limit for $E_\theta{}^*$ consider a real variable ζ which is restricted to values > 1. For any random variable u and any relation R we shall denote by $E(u \mid R)$ the conditional expected value of u under the restriction that R holds. Let $P_\theta(\zeta)$ denote the probability that $e^{h(\theta)Z_{n-1}} < \zeta B^{h(\theta)}$. Then we have

$$(\text{A:21}) \qquad E_\theta{}^* = \int_1^\infty \left[\zeta B^{h(\theta)} E_\theta \left(e^{h(\theta)z} \mid e^{h(\theta)z} \leqq \frac{1}{\zeta} \right) \right] dP_\theta(\zeta)$$

Hence, a lower bound of $E_\theta{}^*$ is given by

$$(\text{A:22}) \qquad B^{h(\theta)} \left[\underset{\zeta}{\text{g.l.b.}} \; \zeta E_\theta \left(e^{h(\theta)z} \mid e^{h(\theta)z} \leqq \frac{1}{\zeta} \right) \right]$$

where the symbol g.l.b. stands for greatest lower bound with respect to ζ. Since $B^{h(\theta)}$ is an upper bound of $E_\theta{}^*$, we obtain the limits

$$(\text{A:23}) \quad B^{h(\theta)} \left[\underset{\zeta}{\text{g.l.b.}} \; \zeta E_\theta \left(e^{h(\theta)z} \mid e^{h(\theta)z} \leqq \frac{1}{\zeta} \right) \right] \leqq E_\theta{}^* \leqq B^{h(\theta)}$$
$$[h(\theta) > 0]$$

To derive limits for $E_\theta{}^{**}$ consider a real variable ρ which is restricted to values > 0 and < 1. Let $Q(\rho)$ denote the probability that $e^{h(\theta)Z_{n-1}} < \rho A^{h(\theta)}$. Then we obtain

$$(\text{A:24}) \qquad E_\theta{}^{**} = \int_0^1 \left[\rho A^{h(\theta)} E_\theta \left(e^{h(\theta)z} \mid e^{h(\theta)z} \geqq \frac{1}{\rho} \right) \right] dQ(\rho)$$

Hence an upper bound of $E_\theta{}^{**}$ is given by

$$(\text{A:25}) \qquad A^{h(\theta)} \left[\underset{\rho}{\text{l.u.b.}} \; \rho E_\theta \left(e^{h(\theta)z} \mid e^{h(\theta)z} \geqq \frac{1}{\rho} \right) \right]$$

Since $A^{h(\theta)}$ is a lower bound of $E_\theta{}^{**}$, we obtain the following limits for $E_\theta{}^{**}$:

$$(\text{A:26}) \quad A^{h(\theta)} \leqq E_\theta{}^{**} \leqq A^{h(\theta)} \left[\underset{\rho}{\text{l.u.b.}} \; \rho E_\theta \left(e^{h(\theta)z} \mid e^{h(\theta)z} \geqq \frac{1}{\rho} \right) \right]$$
$$[h(\theta) > 0]$$

Putting

$$\text{(A:27)} \qquad \operatorname*{g.l.b.}_{\zeta} \zeta E_\theta \left(e^{h(\theta)z} \mid e^{h(\theta)z} \leqq \frac{1}{\zeta} \right) = \eta_\theta$$

and

$$\text{(A:28)} \qquad \operatorname*{l.u.b.}_{\rho} \rho E_\theta \left(e^{h(\theta)z} \mid e^{h(\theta)z} \geqq \frac{1}{\rho} \right) = \delta_\theta$$

inequalities (A:23) and (A:26) can be written as

$$\text{(A:29)} \qquad B^{h(\theta)} \eta_\theta \leqq E_\theta{}^* \leqq B^{h(\theta)}$$

and

$$\text{(A:30)} \qquad A^{h(\theta)} \leqq E_\theta{}^{**} \leqq A^{h(\theta)} \delta_\theta$$

Since $B < 1$ and $A > 1$,[4] we see $E_\theta{}^* < 1$ and $E_\theta{}^{**} > 1$ if $h(\theta) > 0$. From this and relations (A:19), (A:29), and (A:30), it follows that

$$\text{(A:31)} \qquad \frac{A^{h(\theta)} - 1}{A^{h(\theta)} - \eta_\theta B^{h(\theta)}} \leqq L(\theta) \leqq \frac{\delta_\theta A^{h(\theta)} - 1}{\delta_\theta A^{h(\theta)} - B^{h(\theta)}}$$

where $h(\theta) > 0$.

If $h(\theta) < 0$, limits for $L(\theta)$ can be obtained as follows. Let $z' = -z$, $A' = 1/B$ and $B' = 1/A$. Consider the sequential test S' defined as follows. Continue taking observations as long as $\log B' < z'_1 + \cdots + z'_m < \log A'$. Terminate the process with one or the other decision, depending on whether $z'_1 + \cdots + z'_m \leqq \log B'$ or $\geqq \log A'$. We shall let $L'(\theta)$ be the probability that at the termination of the process the cumulative sum $z'_1 + \cdots + z'_m$ is less than or equal to $\log B'$. Then $L'(\theta) = 1 - L(\theta)$. Furthermore, we shall denote the quantities $h(\theta)$, η_θ, δ_θ corresponding to the test S' by $h'(\theta)$, η'_θ, and δ'_θ, respectively. We can apply (A:31) to the test S', since $h'(\theta) = -h(\theta) > 0$. Thus, we obtain

$$\text{(A:32)} \qquad \frac{A'^{h'(\theta)} - 1}{A'^{h'(\theta)} - \eta'_\theta B'^{h'(\theta)}} \leqq L'(\theta) \leqq \frac{\delta'_\theta A'^{h'(\theta)} - 1}{\delta'_\theta A'^{h'(\theta)} - B'^{h'(\theta)}}$$

where $h'(\theta) > 0$. Since η_θ and δ_θ depend only on the distribution of $h(\theta)z$, and since $h'(\theta)z' = h(\theta)z$, we have $\eta'_\theta = \eta_\theta$ and $\delta'_\theta = \delta_\theta$. Substituting, in (A:32), δ_θ for δ'_θ, η_θ for η'_θ, $1/B$ for A', $1/A$ for B', $-h(\theta)$ for $h'(\theta)$, and $1 - L(\theta)$ for $L'(\theta)$, we obtain

---

[4] We have assumed that $B < A$. Since we let $B = \beta/(1-\alpha)$ and $A = (1-\beta)/\alpha$, we must have $\beta/(1-\alpha) < (1-\beta)/\alpha$. Multiplying this inequality by $\alpha(1-\alpha)$, we obtain $\alpha\beta < 1 - \alpha - \beta + \alpha\beta$, i.e., $0 < 1 - \alpha - \beta$. Hence $\beta < 1 - \alpha$ and $1 - \beta > \alpha$, and therefore $B < 1$ and $A > 1$.

(A:33) $$\frac{B^{h(\theta)} - 1}{B^{h(\theta)} - \eta_\theta A^{h(\theta)}} \leqq 1 - L(\theta) \leqq \frac{\delta_\theta B^{h(\theta)} - 1}{\delta_\theta B^{h(\theta)} - A^{h(\theta)}}$$

where $h(\theta) < 0$.   Hence

(A:34) $$\frac{1 - A^{h(\theta)}}{\delta_\theta B^{h(\theta)} - A^{h(\theta)}} \leqq L(\theta) \leqq \frac{1 - \eta_\theta A^{h(\theta)}}{B^{h(\theta)} - \eta_\theta A^{h(\theta)}}$$

where $h(\theta) < 0$.

We can summarize our results as follows.   If $h(\theta) > 0$, limits for $L(\theta)$ are given in (A:31).   If $h(\theta) < 0$, limits for $L(\theta)$ are given in (A:34).   The quantities δ_θ and η_θ are defined in (A:27) and (A:28).

In Sections A.2.4 and A.2.5 we shall calculate the values of δ_θ and η_θ for binomial and normal distributions.   If the limits of $L(\theta)$ as given in (A:31) and (A:34) are too far apart, it may be desirable to determine the exact value of $L(\theta)$, or at least to find a closer approximation to $L(\theta)$ than that given in (A:31) and (A:34).   A method of dealing with this problem is described in Section A.4.   There the exact value of $L(\theta)$ is derived when z can take only a finite number of integral multiples of a constant d.   If z does not have this property, arbitrarily fine approximations to the value of $L(\theta)$ can be obtained, since the distribution of z can be approximated to any desired degree by a discrete distribution of the type mentioned above if the constant d is chosen sufficiently small.

A.2.4   Calculation of δ_θ and η_θ for Binomial Distributions

Let X be a random variable which can take only the values 0 and 1. Let p_i be the probability that $X = 1$ when H_i is true $(i = 0, 1)$.   Let H be the hypothesis that p is the probability that $X = 1$.   Denote $1 - p$ by q and $1 - p_i$ by q_i $(i = 0, 1)$.   The distribution $f(x, p)$ of x is given as follows: $f(1, p) = p$ and $f(0, p) = q$.   It can be assumed without loss of generality that $p_1 > p_0$.   The moment generating function of $z = \log \dfrac{f(x, p_1)}{f(x, p_0)}$ is given by

$$\phi(t) = E_p(e^{zt}) = E_p\left[\frac{f(x, p_1)}{f(x, p_0)}\right]^t = p\left(\frac{p_1}{p_0}\right)^t + q\left(\frac{q_1}{q_0}\right)^t$$

Let $h(p) \neq 0$ be the value of t for which $\phi(t) = 1$, i.e.,

$$p\left(\frac{p_1}{p_0}\right)^{h(p)} + q\left(\frac{q_1}{q_0}\right)^{h(p)} = 1$$

First we consider the case when $h(p) > 0$. It is clear that $e^{zh(p)} = \left[\dfrac{f(x, p_1)}{f(x, p_0)}\right]^{h(p)} > 1$ implies that $x = 1$. Hence $e^{zh(p)} > 1$ implies that $e^{zh(p)} = \left[\dfrac{f(1, p_1)}{f(1, p_0)}\right]^{h(p)} = \left(\dfrac{p_1}{p_0}\right)^{h(p)}$. From this and the definition of δ_p given in (A:28) it follows that

$$(\text{A:35}) \qquad\qquad \delta_p = \left(\frac{p_1}{p_0}\right)^{h(p)}$$

where $h(p) > 0$. Similarly, the inequality $e^{zh(p)} < 1$ implies that $e^{zh(p)} = (q_1/q_0)^{h(p)}$. From this and the definition of η_p given in (A:27) it follows that

$$(\text{A:36}) \qquad\qquad \eta_p = \left(\frac{q_1}{q_0}\right)^{h(p)}$$

where $h(p) > 0$.

If $h(p) < 0$, it can be shown in a similar way that

$$(\text{A:37}) \qquad\qquad \delta_p = \left(\frac{q_1}{q_0}\right)^{h(p)}$$

where $h(p) < 0$, and

$$(\text{A:38}) \qquad\qquad \eta_p = \left(\frac{p_1}{p_0}\right)^{h(p)}$$

where $h(p) < 0$.

A.2.5   Calculation of δ_θ and η_θ for Normal Distributions

We shall now assume that X is normally distributed with unknown mean θ and known variance σ^2. We can assume without loss of generality that $\sigma = 1$, since this can always be achieved by multiplication by a proportionality factor. Then

$$(\text{A:39}) \qquad f(x, \theta_i) = \frac{1}{\sqrt{2\pi}} e^{-\frac{1}{2}(x-\theta_i)^2} \qquad (i = 0, 1)$$

and

$$(\text{A:40}) \qquad f(x, \theta) = \frac{1}{\sqrt{2\pi}} e^{-\frac{1}{2}(x-\theta)^2}$$

We can assume without loss of generality that $\theta_0 = -\Delta$ and $\theta_1 = \Delta$ where $\Delta > 0$, since this can always be achieved by a translation. Then

$$(\text{A:41}) \qquad\qquad z = \log\frac{f(x, \theta_1)}{f(x, \theta_0)} = 2\Delta x.$$

The moment generating function of z is given by

$$(A:42) \qquad E_\theta(e^{zt}) = e^{2\Delta\theta t + 2\Delta^2 t^2}$$

Hence

$$(A:43) \qquad h(\theta) = -\frac{\theta}{\Delta}$$

Substituting this value of $h(\theta)$ in (A:27) and (A:28) we obtain

$$(A:44) \qquad \delta_\theta = \underset{\rho}{\text{l.u.b.}}\ \rho E_\theta\left(e^{-2\theta x}\ \middle|\ e^{-2\theta x} \geqq \frac{1}{\rho}\right)$$

and

$$(A:45) \qquad \eta_\theta = \underset{\zeta}{\text{g.l.b.}}\ \zeta E_\theta\left(e^{-2\theta x}\ \middle|\ e^{-2\theta x} \leqq \frac{1}{\zeta}\right)$$

For any relation R let $P_\theta{}^*(R)$ denote the probability that the relation R holds under the assumption that the distribution of x is normal with mean θ and variance unity. Furthermore, let $P_\theta{}^{**}(R)$ denote the probability that R holds if the distribution of x is normal with mean $-\theta$ and variance unity. Since $e^{-2\theta x}$ is equal to the ratio of the normal probability density function with mean $-\theta$ and variance unity to the normal probability density function with mean θ and variance unity, we see that

$$(A:46) \qquad E_\theta\left(e^{-2\theta x}\ \middle|\ e^{-2\theta x} \geqq \frac{1}{\rho}\right) = \frac{P_\theta{}^{**}\left(e^{-2\theta x} \geqq \frac{1}{\rho}\right)}{P_\theta{}^*\left(e^{-2\theta x} \geqq \frac{1}{\rho}\right)}$$

and

$$(A:47) \qquad E_\theta\left(e^{-2\theta x}\ \middle|\ e^{-2\theta x} \leqq \frac{1}{\zeta}\right) = \frac{P_\theta{}^{**}\left(e^{-2\theta x} \leqq \frac{1}{\zeta}\right)}{P_\theta{}^*\left(e^{-2\theta x} \leqq \frac{1}{\zeta}\right)}$$

It can easily be verified that the right-hand members of (A:46) and (A:47) have the same values for $\theta = \lambda$ as for $\theta = -\lambda$. Thus, δ_θ and η_θ also have the same values for $\theta = \lambda$ as for $\theta = -\lambda$. It will therefore be sufficient to compute δ_θ and η_θ for negative values of θ. Let $\theta = -\lambda$ where $\lambda > 0$. First we show that $\eta_\theta = 1/\delta_\theta$. Clearly,

$$(A:48) \qquad \frac{\zeta P_\theta{}^{**}\left(e^{2\lambda x} \leqq \frac{1}{\zeta}\right)}{P_\theta{}^*\left(e^{2\lambda x} \leqq \frac{1}{\zeta}\right)} = \frac{\zeta P_\theta{}^{**}(e^{-2\lambda x} \geqq \zeta)}{P_\theta{}^*(e^{-2\lambda x} \geqq \zeta)} \qquad (1 \leqq \zeta < \infty)$$

Letting $\zeta = (1/\rho)$ $(0 < \rho \leqq 1)$ in (A:48) gives

(A:49)
$$\frac{\zeta P_\theta{}^{**}\left(e^{2\lambda x} \leqq \dfrac{1}{\zeta}\right)}{P_\theta{}^{*}\left(e^{2\lambda x} \leqq \dfrac{1}{\zeta}\right)} = \frac{P_\theta{}^{**}\left(e^{-2\lambda x} \geqq \dfrac{1}{\rho}\right)}{\rho P_\theta{}^{*}\left(e^{-2\lambda x} \geqq \dfrac{1}{\rho}\right)}$$

Hence

(A:50)
$$\eta_\theta = \underset{\zeta}{\text{g.l.b.}}\left[\frac{\zeta P_\theta{}^{**}\left(e^{2\lambda x} \leqq \dfrac{1}{\zeta}\right)}{P_\theta{}^{*}\left(e^{2\lambda x} \leqq \dfrac{1}{\zeta}\right)}\right] = \frac{1}{\underset{\rho}{\text{l.u.b.}}\left[\dfrac{\rho P_\theta{}^{*}\left(e^{-2\lambda x} \geqq \dfrac{1}{\rho}\right)}{P_\theta{}^{**}\left(e^{-2\lambda x} \geqq \dfrac{1}{\rho}\right)}\right]}$$

Because of the symmetry of the normal distribution, it is easily seen that

$$\underset{\rho}{\text{l.u.b.}}\left[\frac{\rho P_\theta{}^{*}\left(e^{-2\lambda x} \geqq \dfrac{1}{\rho}\right)}{P_\theta{}^{**}\left(e^{-2\lambda x} \geqq \dfrac{1}{\rho}\right)}\right] = \text{l.u.b.}\left[\frac{\rho P_\theta{}^{**}\left(e^{2\lambda x} \geqq \dfrac{1}{\rho}\right)}{P_\theta{}^{*}\left(e^{2\lambda x} \geqq \dfrac{1}{\rho}\right)}\right] = \delta_\theta$$

Hence

(A:51)
$$\eta_\theta = \frac{1}{\delta_\theta}$$

Now we shall calculate the value of δ_θ. Let $G(x)$ denote $\dfrac{1}{\sqrt{2\pi}}\displaystyle\int_x^\infty e^{-\frac{t^2}{2}}\,dt$. Then

$$P_\theta{}^{**}\left(e^{2\lambda x} \geqq \frac{1}{\rho}\right) = P_\theta{}^{**}\left(2\lambda x \geqq \log\frac{1}{\rho}\right) = P_\theta{}^{**}\left(x \geqq \frac{1}{2\lambda}\log\frac{1}{\rho}\right)$$

$$= G\left(\frac{1}{2\lambda}\log\frac{1}{o} - \lambda\right)$$

Similarly

$$P_\theta{}^{*}\left(e^{2\lambda x} \geqq \frac{1}{\rho}\right) = P_\theta{}^{*}\left(x \geqq \frac{1}{2\lambda}\log\frac{1}{\rho}\right) = G\left(\frac{1}{2\lambda}\log\frac{1}{\rho} + \lambda\right)$$

Let u denote $(1/2\lambda) \log (1/\rho)$. Since ρ can vary from 0 to 1, u can take any value from 0 to ∞. Since $\rho = e^{-2\lambda u}$, we have

$$(A{:}52) \quad \delta_\theta = \text{l.u.b.} \left[\frac{\rho P_\theta^{**} \left(e^{2\lambda x} \geq \dfrac{1}{\rho} \right)}{P_\theta^* \left(e^{2\lambda x} \geq \dfrac{1}{\rho} \right)} \right] = \text{l.u.b.}_u \left(e^{-2\lambda u} \frac{G(u - \lambda)}{G(u + \lambda)} \right)$$

$$(0 \leq u \leq \infty)$$

We shall prove that

$$(A{:}53) \qquad\qquad \chi(u) = e^{-2u\lambda} \frac{G(u - \lambda)}{G(u + \lambda)}$$

s a monotonically decreasing function of u and consequently has a maximum at $u = 0$. For this purpose it suffices to show that the derivative of $\log \chi(u)$ is never positive. Now

$$(A{:}54) \qquad \log \chi(u) = \log G(u - \lambda) - \log G(u + \lambda) - 2\lambda u$$

Let $\Phi(x)$ denote $\dfrac{1}{\sqrt{2\pi}} e^{-\frac{1}{2}x^2}$. Since $\dfrac{d}{du} G(u) = -\Phi(u)$, it follows from (A:54) that

$$(A{:}55) \quad \cdot \quad \frac{d}{du} \log \chi(u) = - \frac{\Phi(u - \lambda)}{G(u - \lambda)} + \frac{\Phi(u + \lambda)}{G(u + \lambda)} - 2\lambda$$

It follows from the mean value theorem that the right-hand side of (A:55) is never positive if $\dfrac{d}{du} \left[\dfrac{\Phi(u)}{G(u)} \right]$ is equal to or less than 1 for all values of u. Thus, we need merely to show that

$$(A{:}56) \quad \frac{d}{du} \left[\frac{\Phi(u)}{G(u)} \right] = \frac{\Phi'(u)G(u) - G'(u)\Phi(u)}{G^2(u)} = \frac{\Phi'(u)G(u) + \Phi^2(u)}{G^2(u)}$$

$$= \frac{\Phi^2(u)}{G^2(u)} - u \frac{\Phi(u)}{G(u)} \leq 1$$

Let y denote $\dfrac{\Phi(u)}{G(u)}$. The roots of the equation $y^2 - uy - 1 = 0$ are

$$y = \frac{u \pm \sqrt{u^2 + 4}}{2}$$

Hence the inequality $y^2 - uy - 1 \leq 0$ holds if, and only if,

$$\frac{u - \sqrt{u^2 + 4}}{2} \leq y \leq \frac{u + \sqrt{u^2 + 4}}{2}$$

Since y cannot be negative, this inequality is equivalent to

$$(A:57) \qquad \frac{\Phi(u)}{G(u)} = y \leqq \frac{u + \sqrt{u^2 + 4}}{2}$$

Thus we merely have to prove (A:57). We shall show that (A:57) holds for all real values of u. Birnbaum [5] has shown that for $u > 0$

$$(A:58) \qquad \frac{\sqrt{u^2 + 4} - u}{2} \, \Phi(u) \leqq G(u)$$

Hence

$$(A:59) \qquad \frac{\Phi(u)}{G(u)} \leqq \frac{2}{\sqrt{u^2 + 4} - u} = \frac{\sqrt{u^2 + 4} + u}{2} \qquad (u > 0)$$

which proves (A:57) for $u > 0$. Now we prove (A:57) for $u < 0$. Let $u = -v$ where $v > 0$. Then it follows from (A:59) that

$$(A:60) \qquad \frac{\Phi(v)}{G(v)} \leqq \frac{2}{\sqrt{4 + v^2} - v}$$

Taking reciprocals, we obtain, from (A:60),

$$(A:61) \qquad \frac{G(v)}{\Phi(v)} \geqq \frac{\sqrt{4 + v^2} - v}{2}$$

Since

$$\frac{G(u)}{\Phi(u)} \geqq \frac{G(v) + 2v\Phi(v)}{\Phi(v)} = \frac{G(v)}{\Phi(v)} + 2v$$

we obtain, from (A:61),

$$(A:62) \qquad \frac{G(u)}{\Phi(u)} \geqq \frac{\sqrt{v^2 + 4} + 3v}{2} \geqq \frac{\sqrt{v^2 + 4} + v}{2}$$

Taking reciprocals, we obtain

$$\frac{\Phi(u)}{G(u)} \leqq \frac{2}{\sqrt{v^2 + 4} + v} = \frac{\sqrt{v^2 + 4} - v}{2} = \frac{\sqrt{u^2 + 4} + u}{2}$$

Hence (A:57) is proved for all values of u and consequently δ_θ is equal to the value of the expression (A:53) if we substitute 0 for u. Thus

$$(A:63) \qquad \delta_\theta = \frac{G(-\lambda)}{G(\lambda)} \qquad (\lambda = |\theta|)$$

[5] Z. W. Birnbaum, "An Inequality for Mills' Ratio," *The Annals of Mathematical Statistics*, Vol. XIII (1942).

Formula (A:63) has been derived for the case in which $\theta_0 = -\Delta$, $\theta_1 = \Delta$, and $\sigma = 1$. It can easily be seen that for general values θ_0, θ_1, and σ we have

$$(\text{A:64}) \qquad\qquad \delta_\theta = \frac{G(-\lambda)}{G(\lambda)}$$

where $\lambda = \dfrac{1}{\sigma}\left| \theta - \dfrac{\theta_0 + \theta_1}{2} \right|$.

A.3 UPPER AND LOWER LIMITS FOR THE ASN FUNCTION OF A SEQUENTIAL PROBABILITY RATIO TEST

A.3.1 Derivation of General Formulas for Upper and Lower Limits

As before, let

$$z = \log\frac{f(x, \theta_1)}{f(x, \theta_0)}, \quad z_i = \log\frac{f(x_i, \theta_1)}{f(x_i, \theta_0)} \qquad (i = 1, 2, \cdots, \text{ad inf.})$$

and let n be the number of observations required by the sequential test, i.e., n is the smallest integer for which $Z_n = z_1 + \cdots + z_n$ is either $\geq \log A$ or $\leq \log B$. To determine the expected value $E(n)$ of n under the hypothesis H that θ is the true value of the parameter, we shall consider a fixed positive integer N. The sum $Z_N = z_1 + \cdots + z_N$ can be split in two parts as follows:

$$(\text{A:65}) \qquad\qquad Z_N = Z_n + Z'_n$$

where $Z'_n = z_{n+1} + \cdots + z_N$ if $n \leq N$ and $Z'_n = Z_N - Z_n$ if $n > N$. Taking expected values on both sides of (A:65) we obtain

$$NE_\theta(z) = E_\theta(Z_n + Z'_n)$$

Let P_N denote the probability that $n \leq N$. Then

$$E_\theta(Z_n + Z'_n) = P_N E_{\theta N}(Z_n + Z'_n) + (1 - P_N)E_{\theta N}*(Z_N)$$

where the operator $E_{\theta N}$ means conditional expected value when $n \leq N$, and $E_{\theta N}*$ means conditional expected value when $n > N$.

Since Z_N lies between $\log B$ and $\log A$ when $n > N$, and since $\lim (1 - P_N) = 0$, we obtain from the last two equations

$$(\text{A:66}) \qquad\qquad \lim_{N=\infty} [NE_\theta(z) - P_N E_{\theta N}(Z_n + Z'_n)] = 0$$

For any given value of $n < N$, the variates $z_{n+1}, \cdots, z_N$ are independently distributed, each having the same distribution as z. Hence, we have

$$E_{\theta N}(Z'_n) = E_{\theta N}(N - n)E_\theta(z) = -E_{\theta N}(n)E_\theta(z) + NE_\theta(z)$$

From this and (A:66) we obtain, since $\lim_{N=\infty} (1 - P_N)N = 0,$[1]

(A:67) $$\lim_{N=\infty} [P_N E_{\theta N}(n)E_\theta(z) - P_N E_{\theta N}(Z_n)] = 0$$

Since

$$\lim_{N=\infty} P_N E_{\theta N}(n) = E_\theta(n) \qquad \text{and} \qquad \lim_{N=\infty} P_N E_{\theta N}(Z_n) = E_\theta(Z_n)$$

equation (A:67) gives

(A:68) $$E_\theta(Z_n) = E_\theta(n)E_\theta(z)$$

Hence

(A:69) $$E_\theta(n) = \frac{E_\theta(Z_n)}{E_\theta(z)}$$

if $E_\theta(z) \neq 0$. Let $E_\theta{}^*(Z_n)$ be the conditional expected value of Z_n under the restriction that the sequential analysis leads to the acceptance of H_0, i.e., that $Z_n \leqq \log B$. Similarly, let $E_\theta{}^{**}(Z_n)$ be the conditional expected value of Z_n under the restriction that H_1 is accepted, i.e., that $Z_n \geqq \log A$. Since $L(\theta)$ is the probability that $Z_n \leqq \log B$, and $1 - L(\theta)$ is the probability that $Z_n \geqq \log A$, we have

(A:70) $$E_\theta(Z_n) = [L(\theta)]E_\theta{}^*(Z_n) + [1 - L(\theta)]E_\theta{}^{**}(Z_n)$$

From (A:69) and (A:70) we obtain

(A:71) $$E_\theta(n) = \frac{[L(\theta)]E_\theta{}^*(Z_n) + [1 - L(\theta)]E_\theta{}^{**}(Z_n)}{E_\theta(z)}$$

The exact value of $E_\theta(Z_n)$, and therefore also the exact value of $E_\theta(n)$, can be computed if z can take only integral multiples of a constant d, since in this case the exact probability distribution of Z_n was obtained (see Section A.4). If z does not satisfy the above restriction, it is still possible to obtain arbitrarily fine approximations to the value

[1] C. Stein has shown, in "A Note on Cumulative Sums," *The Annals of Mathematical Statistics*, Vol. 17 (1946), that all moments of n must be finite. This implies that $\lim_{N=\infty} (1 - P_N)N^k = 0$ for any positive integer k.

of $E_\theta(Z_n)$, since the distribution of z can be approximated to any desired degree by a discrete distribution of the type mentioned above provided the constant d is chosen sufficiently small.

If both $|E(z)|$ and the standard deviation of z are small, $E_\theta^*(Z_n)$ is very nearly equal to $\log B$ and $E_\theta^{**}(Z_n)$ is very nearly equal to $\log A$. Hence in this case we can write

$$(A{:}72) \qquad E_\theta(n) \sim \frac{[L(\theta)]\log B + [1 - L(\theta)]\log A}{E_\theta(z)}$$

This is the same approximation formula as given in (3:57).

To judge the goodness of the approximation given in (A:72) we shall derive lower and upper limits for $E_\theta(n)$ by deriving lower and upper limits for $E_\theta^*(Z_n)$ and $E_\theta^{**}(Z_n)$. Let r be a non-negative variable and let

$$(A{:}73) \qquad \xi_\theta = \operatorname*{Max}_r E_\theta(z - r \,|\, z \geqq r) \qquad (r \geqq 0)$$

and

$$(A{:}74) \qquad \xi'_\theta = \operatorname*{Min}_r E_\theta(z + r \,|\, z + r \leqq 0) \qquad (r \geqq 0)$$

It is easy to see that

$$(A{:}75) \qquad \log A \leqq E_\theta^{**}(Z_n) \leqq \log A + \xi_\theta$$

and

$$(A{:}76) \qquad \log B + \xi'_\theta \leqq E_\theta^*(Z_n) \leqq \log B$$

We obtain from (A:71), (A:75), and (A:76)

$$(A{:}77) \qquad \frac{L(\theta)(\log B + \xi'_\theta) + [1 - L(\theta)]\log A}{E_\theta(z)} \leqq E_\theta(n)$$

$$\leqq \frac{[L(\theta)]\log B + [1 - L(\theta)](\log A + \xi_\theta)}{E_\theta(z)} \quad \text{if } E_\theta(z) > 0$$

and

$$(A{:}78) \qquad \frac{[L(\theta)]\log B + [1 - L(\theta)](\log A + \xi_\theta)}{E_\theta(z)} \leqq E_\theta(n)$$

$$\leqq \frac{L(\theta)(\log B + \xi'_\theta) + [1 - L(\theta)]\log A}{E_\theta(z)} \quad \text{if } E_\theta(z) < 0$$

The limits given in (A:77) and (A:78) will generally be close to each other for values $\theta \leqq \theta_0$ and $\theta \geqq \theta_1$. However, for values θ between

θ_0 and θ_1 the difference between the upper and lower limits may become very large, since $E_\theta(z)$ may be near (or equal to) 0 for such values θ. In fact, we have seen that $E_{\theta_0}(z) < 0$ and $E_{\theta_1}(z) > 0$. Hence, if $E_\theta(z)$ is a continuous function of θ, there will be a value θ' between θ_0 and θ_1 such that $E_{\theta'}(z) = 0$. For $\theta = \theta'$ or for values θ very near θ' the limits given in (A:77) and (A:78) are of no practical value, since they are far apart.

We shall now derive limits for $E_\theta(n)$ which can be used for values θ in the neighborhood of θ'.[2] For this purpose, we shall expand $e^{h(\theta)Z_n}$ in a Taylor series as follows:

$$(A:79) \qquad e^{h(\theta)Z_n} = 1 + h(\theta)Z_n + \tfrac{1}{2}[h(\theta)]^2 Z_n^2 + \tfrac{1}{6}[h(\theta)]^3 Z_n^3 e^\lambda$$

where λ is some value between 0 and $h(\theta)Z_n$. From (A:17) and (A:79) we obtain

$$(A:80) \qquad h(\theta)E_\theta(Z_n) = -\tfrac{1}{2}[h(\theta)]^2 E_\theta(Z_n^2) - \tfrac{1}{6}[h(\theta)]^3 E_\theta(Z_n^3 e^\lambda)$$

From this and (A:69) it follows that

$$(A:81) \qquad E_\theta(n) = -\frac{h(\theta)}{2E_\theta(z)} E_\theta(Z_n^2) - \frac{[h(\theta)]^2}{6E_\theta(z)} E_\theta(Z_n^3 e^\lambda)$$

Thus, upper and lower limits for $E_\theta(n)$ can be obtained by deriving upper and lower limits for $E_\theta(Z_n^2)$ and $E_\theta(Z_n^3 e^\lambda)$. To derive limits for $E_\theta(Z_n^2)$, we write

$$(A:82) \qquad E_\theta(Z_n^2) = L(\theta)E_\theta^*(Z_n^2) + [1 - L(\theta)]E_\theta^{**}(Z_n^2)$$

where the operator E^* stands for conditional expected value when $Z_n \leq \log B$, and E^{**} stands for conditional expected value when $Z_n \geq \log A$. Let ϵ' denote $Z_n - \log B$ and ϵ'' denote $Z_n - \log A$. Then

$$(A:83) \qquad E_\theta^*(Z_n^2) = (\log B)^2 + 2(\log B)E_\theta^*(\epsilon') + E_\theta^*(\epsilon'^2)$$

and

$$(A:84) \quad E_\theta^{**}(Z_n^2) = (\log A)^2 + 2(\log A)E_\theta^{**}(\epsilon'') + E_\theta^{**}(\epsilon''^2)$$

Since $E_\theta^*(\epsilon'^2) \geq 0$ and $(\log B)E_\theta^*(\epsilon') \geq 0$, we obtain, from (A:83),

$$(A:85) \qquad\qquad (\log B)^2 \leq E_\theta^*(Z_n)^2$$

[2] See also the author's paper, "Some Improvements in Setting Limits for the Expected Number of Observations Required by a Sequential Probability Ratio Test," *The Annals of Mathematical Statistics*, Vol. 17 (1946).

The quantity ξ'_θ given in (A:74) is a lower bound for $E_\theta^*(\epsilon')$. Since $\log B < 0$, $(\log B)\xi'_\theta$ is an upper bound for $(\log B)E_\theta^*(\epsilon')$. An upper bound for $E_\theta^*(\epsilon'^2)$ is given by

$$(A:86) \qquad \zeta'_\theta = \operatorname*{Max}_r E_\theta[(z + r)^2 \mid z + r \leq 0] \qquad (r \geq 0)$$

Hence

$$(A:87) \qquad E_\theta^*(Z_n{}^2) \leq (\log B)^2 + 2(\log B)\xi'_\theta + \zeta'_\theta$$

Thus we obtain the limits

$$(A:88) \qquad (\log B)^2 \leq E_\theta^*(Z_n{}^2) \leq (\log B)^2 + 2(\log B)\xi'_\theta + \zeta'_\theta$$

In a similar way, the following limits can be derived for $E_\theta^{**}(Z_n{}^2)$:

$$(A:89) \qquad (\log A)^2 \leq E_\theta^{**}(Z_n{}^2) \leq (\log A)^2 + 2(\log A)\xi_\theta + \zeta_\theta$$

where ξ_θ is given in (A:73) and

$$(A:90) \qquad \zeta_\theta = \operatorname*{Max}_r E_\theta[(z - r)^2 \mid z \geq r] \qquad (r \geq 0)$$

If we denote by $L'(\theta)$ the lower limit and by $L''(\theta)$ the upper limit of $L(\theta)$ given in (A:31) [(A:34) when $h(\theta) < 0$], we obtain from (A:82), (A:88), and (A:89) the following limits for $E_\theta(Z_n{}^2)$:

$$(A:91) \quad L'(\theta)(\log B)^2 + [1 - L''(\theta)](\log A)^2 \leq E_\theta(Z_n{}^2)$$
$$\leq L''(\theta)[(\log B)^2 + 2(\log B)\xi'_\theta + \zeta'_\theta] +$$
$$[1 - L'(\theta)][(\log A)^2 + 2(\log A)\xi_\theta + \zeta_\theta]$$

Using a similar method, one can also derive upper and lower limits for $E_\theta(Z_n{}^3 e^\lambda)$ without any difficulty. We shall, however, not derive such limits here, since we are interested in obtaining limits for $E_\theta(n)$ when θ is near θ' and since, for such values of θ, the second term in the right-hand member of (A:81) is negligible. We shall show that, if $h(\theta)$, $E_\theta(z)$, and $E_\theta(z^2)$ are continuous functions of θ, the factor $[h(\theta)]^2/[E_\theta(z)]$ in that term converges to 0 as $\theta \to \theta'$. It follows from the discussion given in Section A.2.1 that $\lim_{\theta = \theta'} h(\theta) = 0$. Since

$$(A:92) \quad E_\theta(e^{h(\theta)z}) = E_\theta\left\{1 + h(\theta)z + \frac{[h(\theta)]^2}{2!}z^2 + \frac{[h(\theta)]^3}{3!}z^3 e^{uh(\theta)z}\right\} = 1$$
$$(0 \leq u \leq 1)$$

we obtain, when $h(\theta) \neq 0$,

$$(A:93) \qquad E_\theta\left\{z + \frac{h(\theta)}{2!}z^2 + \frac{[h(\theta)]^2}{3!}z^3 e^{uh(\theta)z}\right\} = 0$$

Thus

(A:94) $$\frac{E_\theta(z)}{h(\theta)} = E_\theta\left[-\frac{1}{2}z^2 - \frac{h(\theta)}{3!}z^3 e^{uh(\theta)z}\right] \qquad [h(\theta) \neq 0]$$

Assuming that $E_\theta(e^{|z|})$ is a bounded function of θ in the neighborhood of θ', we see that $E_\theta(|z|^3 e^{|h(\theta)||z|})$ is also a bounded function of θ in a sufficiently small neighborhood of θ'.[3] Hence, $E_\theta(z^3 e^{uh(\theta)z})$ is also a bounded function of θ in the neighborhood of θ'. From this and (A:94) it follows that

(A:95) $$\lim_{\theta=\theta'} \frac{E_\theta(z)}{h(\theta)} = -\frac{1}{2}E_{\theta'}(z^2) < 0$$

From (A:95) it follows that

(A:96) $$\lim_{\theta=\theta'} \frac{[h(\theta)]^2}{E_\theta(z)} = 0$$

The lower and upper limits for $E_\theta(n)$, based on (A:81), will generally be close to each other for values θ in a small neighborhood of θ'. Thus, when θ is near θ' these limits for $E_\theta(n)$ can be used instead of the limits given in (A:77) and (A:78).

It may be of interest to determine the limiting form of (A:81) when $\theta = \theta'$. If $E_\theta(Z_n^2)$ is a continuous function of θ and $E_\theta(Z_n^3 e^\lambda)$ is a bounded function of θ in the neighborhood of θ', it follows from (A:81), (A:95), and (A:96) that[4]

(A:97) $$E_{\theta'}(n) = \frac{E_{\theta'}(Z_n^2)}{E_{\theta'}(z^2)}$$

The boundedness of $E_\theta(Z_n^3 e^\lambda)$ can be proved if, for $t = \pm 1$, the expected value $\rho E_\theta\left(e^{tz} \mid e^{tz} \geq \frac{1}{\rho}\right)$ is a bounded function of θ and ρ $(0 < \rho < 1)$. Since $\lim_{\theta=\theta'} h(\theta) = 0$, there exists a constant C such that $|Z_n^3 e^\lambda| \leq C e^{|Z_n|}$ for θ in the neighborhood of θ'. Hence, we merely have to show that $E_\theta(e^{|Z_n|})$ is bounded. Since $e^{Z_n} + e^{-Z_n} \geq e^{|Z_n|}$, it is sufficient to show that both $E_\theta(e^{Z_n})$ and $E_\theta(e^{-Z_n})$ are bounded. We have

$$E_\theta(e^{Z_n} \mid Z_n \geq \log A) \leq A \text{ l.u.b.}_\rho \left[\rho E_\theta\left(e^z \mid e^z \geq \frac{1}{\rho}\right)\right]$$

---

[3] This follows from the fact that $|h(\theta)| < 1$ when θ is sufficiently near θ'.

[4] A different method for deriving (A:97) was given in the author's paper, "Differentiation under the Expectation Sign in the Fundamental Identity," *The Annals of Mathematical Statistics*, Vol. 17 (1946).

where $0 < \rho < 1$. Since

$$E_\theta(e^{Z_n} \mid Z_n \leq \log B) \leq B$$

we obtain

$$E_\theta(e^{Z_n}) \leq A \, \text{l.u.b.}_\rho \left[\rho E_\theta \left(e^z \mid e^z \geq \frac{1}{\rho} \right) \right] + B$$

The right-hand member of this equation is bounded, since $\rho E_\theta \left(e^z \mid e^z \geq \dfrac{1}{\rho} \right)$ is bounded by assumption. Hence $E_\theta(e^{Z_n})$ is bounded.

The boundedness of $E_\theta(e^{-Z_n})$ can be shown in a similar way. Upper and lower limits for $E_{\theta'}(n)$ can be obtained from (A:97) by substituting for $E_{\theta'}(Z_n{}^2)$ the upper and lower limits given in (A:91).

We shall now compute an approximate value of $E_{\theta'}(n)$, neglecting the excess of Z_n over the boundaries. Since $\lim\limits_{\theta=\theta'} h(\theta) = 0$, we obtain, from (3:43),

(A:98)
$$L(\theta') \sim \frac{\log A}{\log A - \log B}$$

Hence

$$E_{\theta'}(Z_n{}^2) \sim \frac{\log A}{\log A - \log B} (\log B)^2 + \frac{-\log B}{\log A - \log B} (\log A)^2$$

$$= - \log B \log A$$

Thus an approximate value of $E_{\theta'}(n)$ is given by [5]

(A:99)
$$E_{\theta'}(n) = \frac{E_{\theta'}(Z_n{}^2)}{E_{\theta'}(z^2)} \sim \frac{- \log B \log A}{E_{\theta'}(z^2)}$$

If the OC function $L(\theta)$ of the test is known exactly, close limits for $E_\theta(n)$ can be derived which remain valid over the entire range of θ. We shall indicate briefly the derivation of such limits. Denote by $f_\theta(z)$ the distribution of z when θ is the true value of the parameter. By the distribution of z conjugate to the distribution $f_\theta(z)$ we shall mean the distribution $e^{h(\theta)z} f_\theta(z)$. In important cases, such as for binomial and normal distributions, to any given value θ of the parameter there will correspond a value $\bar\theta$ such that $f_{\bar\theta}(z)$ is conjugate to

[5] W. Allen Wallis obtained this approximation formula independently of the author. It is included in the publication of the Statistical Research Group of Columbia University, *Techniques of Statistical Analysis*, Chapter 17, Section 7.2, McGraw-Hill, New York (1946).

$f_\theta(z)$, i.e., $f_{\bar\theta}(z) = e^{h(\theta)z}f_\theta(z)$. We shall call $\bar\theta$ conjugate to θ. It has been shown elsewhere [6] that

(A:100) $E_\theta{}^*(e^{h(\theta)Z_n}) = \dfrac{L(\bar\theta)}{L(\theta)}$ and $E_\theta{}^{**}(e^{h(\theta)Z_n}) = \dfrac{1 - L(\bar\theta)}{1 - L(\theta)}$

On the other hand,

(A:101) $E_\theta{}^*(e^{h(\theta)Z_n}) = e^{h(\theta)E_\theta{}^*(Z_n)}E_\theta{}^*(e^{h(\theta)[Z_n - E_\theta{}^*(Z_n)]})$

$$= e^{h(\theta)E_\theta{}^*(Z_n)}E_\theta{}^*\left\{1 + \frac{[h(\theta)]^2}{2}[Z_n - E_\theta{}^*(Z_n)]^2 e^v\right\}$$

where v lies between 0 and $h(\theta)[Z_n - E_\theta{}^*(Z_n)]$. Similarly

(A:102) $E_\theta{}^{**}(e^{h(\theta)Z_n})$

$$= e^{h(\theta)E_\theta{}^{**}(Z_n)}E_\theta{}^{**}\left\{1 + \frac{[h(\theta)]^2}{2}[Z_n - E_\theta{}^{**}(Z_n)]^2 e^{v'}\right\}$$

where v' lies between 0 and $h(\theta)[Z_n - E_\theta{}^{**}(Z_n)]$. From (A:100), (A:101), and (A:102) we obtain

(A:103) $\dfrac{E_\theta{}^*(Z_n)}{E_\theta(z)} = \dfrac{1}{h(\theta)E_\theta(z)}\log\dfrac{L(\bar\theta)}{L(\theta)} -$

$$\frac{1}{h(\theta)E_\theta(z)}\log\left(1 + \frac{[h(\theta)]^2}{2}E_\theta{}^*\left\{[Z_n - E_\theta{}^*(Z_n)]^2 e^v\right\}\right)$$

and

(A:104) $\dfrac{E_\theta{}^{**}(Z_n)}{E_\theta(z)} = \dfrac{1}{h(\theta)E_\theta(z)}\log\dfrac{1 - L(\bar\theta)}{1 - L(\theta)} -$

$$\frac{1}{h(\theta)E_\theta(z)}\log\left(1 + \frac{[h(\theta)]^2}{2}E_\theta{}^{**}\left\{[Z_n - E_\theta{}^{**}(Z_n)]^2 e^{v'}\right\}\right)$$

Thus

(A:105) $E_\theta(n)$

$$= \frac{1}{h(\theta)E_\theta(z)}\left\{L(\theta)\log\frac{L(\bar\theta)}{L(\theta)} + [1 - L(\theta)]\log\frac{1 - L(\bar\theta)}{1 - L(\theta)}\right\} + R$$

where

(A:106)

$$R = -\frac{1}{h(\theta)E_\theta(z)}\left[L(\theta)\log\left(1 + \frac{[h(\theta)]^2}{2}E_\theta{}^*\left\{[Z_n - E_\theta{}^*(Z_n)]^2 e^v\right\}\right) + \right.$$

$$\left. [1 - L(\theta)]\log\left(1 + \frac{[h(\theta)]^2}{2}E_\theta{}^{**}\left\{[Z_n - E_\theta{}^{**}(Z_n)]^2 e^{v'}\right\}\right)\right]$$

[6] See, for instance, the author's article on "Some Generalizations of the Theory of Cumulative Sums of Random Variables," *The Annals of Mathematical Statistics*, Vol. XVI (1945).

Since $h(\theta)E_\theta(z) \leq 0$ (see Section A.2.1), we see that $R \geq 0$. Hence a lower bound for $E_\theta(n)$ is obtained by substituting 0 for R in (A:105).

To obtain an upper bound for $E_\theta(n)$ we shall derive an upper bound for R. Clearly

(A:107) $\{(Z_n - \log B) + [E_\theta{}^*(Z_n) - \log B]\}^2 \geq [Z_n - E_\theta{}^*(Z_n)]^2$

whenever $Z_n \leq \log B$. From this and (A:76) we obtain

(A:108) $[(Z_n - \log B) + \xi'_\theta]^2 \geq [Z_n - E_\theta{}^*(Z_n)]^2$

whenever $Z_n \leq \log B$. Similarly, we obtain

(A:109) $[(Z_n - \log A) + \xi_\theta]^2 \geq [Z_n - E_\theta{}^{**}(Z_n)]^2$

whenever $Z_n \geq \log A$, where ξ_θ is given by (A:73). From (A:107), (A:108), and (A:109) it follows that

(A:110) $E_\theta{}^*\{[Z_n - E_\theta{}^*(Z_n)]^2 e^v\}$

$$\leq E_\theta{}^*[(Z_n - \log B + \xi'_\theta)^2 e^{|\,Z_n - \log B + \xi'_\theta\,|\,|\,h(\theta)\,|}]$$

and

(A:111) $E_\theta{}^{**}\{[Z_n - E_\theta{}^{**}(Z_n)]^2 e^{v'}\}$

$$\leq E_\theta{}^{**}[(Z_n - \log A + \xi_\theta)^2 e^{(Z_n - \log A + \xi_\theta)|\,h(\theta)\,|}]$$

Furthermore, we have

(A:112) $E_\theta{}^*[(Z_n - \log B + \xi'_\theta)^2 e^{|\,Z_n - \log B + \xi'_\theta\,|\,|\,h(\theta)\,|}]$

$\leq \underset{r \geq 0}{\text{Max }} E_\theta{}^*[(z + r + \xi'_\theta)^2 e^{|\,z + r + \xi'_\theta\,|\,|\,h(\theta)\,|}\,|\,z + r \leq 0] = \rho'$ (say)

and

(A:113) $E_\theta{}^{**}[(Z_n - \log A + \xi_\theta)^2 e^{(Z_n - \log A + \xi_\theta)|\,h(\theta)\,|}]$

$\leq \underset{r \geq 0}{\text{Max }} E_\theta{}^{**}[(z - r + \xi_\theta)^2 e^{(z - r + \xi_\theta)|\,h(\theta)\,|}\,|\,z - r \geq 0] = \rho''$ (say)

From (A:106) and (A:110) through (A:113) we obtain the following upper bound for R:

(A:114) $R \leq \bar{R} = -\dfrac{1}{h(\theta)E_\theta(z)}\left(L(\theta)\log\left\{1 + \dfrac{[h(\theta)]^2}{2}\rho'\right\} + \right.$

$$\left. [1 - L(\theta)]\log\left\{1 + \dfrac{[h(\theta)]^2}{2}\rho''\right\}\right)$$

An upper limit for $E_\theta(n)$ is obtained by substituting $\bar{R}$ for R in (A:105). The value of $\bar{R}$ will generally be small over the entire range of θ.

A.3.2 Calculation of the Quantities ξ_θ and ξ'_θ for Binomial and Normal Distributions

Let X be a random variable which can take only the values 0 and 1. Let the probability that $X = 1$ be denoted by θ. Then the distribution of x is given by $f(x, \theta)$, where $f(1, \theta) = \theta$ and $f(0, \theta) = 1 - \theta$. Let H_i be the hypothesis that $\theta = \theta_i$ $(i = 0, 1)$. It can be assumed without loss of generality that $\theta_1 > \theta_0$. It is clear that $\log \dfrac{f(x, \theta_1)}{f(x, \theta_0)} > 0$

implies that $x = 1$ and consequently $\log \dfrac{f(x, \theta_1)}{f(x, \theta_0)} = \log \dfrac{f(1, \theta_1)}{f(1, \theta_0)} =$

$\log \dfrac{\theta_1}{\theta_0}$. Hence

$$(A:115) \qquad \xi_\theta = \operatorname*{Max}_r E_\theta(z - r \mid z \geqq r) = \log \frac{\theta_1}{\theta_0}$$

Since $\log \dfrac{f(x, \theta_1)}{f(x, \theta_0)} \leqq 0$ implies that $x = 0$, we have

$$(A:116) \qquad \xi'_\theta = \operatorname*{Min}_r E_\theta(z + r \mid z + r \leqq 0) = \log \frac{1 - \theta_1}{1 - \theta_0}$$

Now we shall calculate the values ξ_θ and ξ'_θ when X is normally distributed with unit variance. Let

$$f(x, \theta_i) = \frac{1}{\sqrt{2\pi}} e^{-\frac{1}{2}(x-\theta_i)^2} \qquad (i = 0, 1 \text{ and } \theta_1 > \theta_0)$$

and

$$f(x, \theta) = \frac{1}{\sqrt{2\pi}} e^{-\frac{1}{2}(x-\theta)^2}$$

We may assume without loss of generality that $\theta_0 = -\Delta$ and $\theta_1 = \Delta$ where $\Delta > 0$, since this can always be achieved by a translation. Then

$$(A:117) \qquad z = \log \frac{f(x, \theta_1)}{f(x, \theta_0)} = 2\Delta x$$

Let $\Phi(x)$ denote $\dfrac{1}{\sqrt{2\pi}} e^{-\frac{x^2}{2}}$ and let $G(x)$ denote $\dfrac{1}{\sqrt{2\pi}} \displaystyle\int_x^\infty e^{-\frac{t^2}{2}} dt$. Let $t = x - \theta$. Then $z = 2\Delta(t + \theta)$ and

$$(A:118) \quad E_\theta(z - r \mid z - r \geqq 0)$$
$$= 2\Delta E_\theta\left(t + \theta - \frac{r}{2\Delta} \mid t + \theta - \frac{r}{2\Delta} \geqq 0\right)$$
$$= \frac{2\Delta}{G(t_0)} \int_{t_0}^\infty (t - t_0)\Phi(t)\, di = \frac{2\Delta}{G(t_0)} [-t_0 G(t_0) + \Phi(t_0)]$$

where

(A:119) $$t_0 = \frac{r}{2\Delta} - \theta$$

In Section A.2.5, equation (A:56), it was proved that $[\Phi(t_0)/G(t_0)]$ $- t_0$ is a monotonically decreasing function of t_0. Hence the maximum of $E_\theta(z - r \mid z - r \geq 0)$ is reached when $r = 0$, and consequently

(A:120) $$\xi_\theta = \frac{2\Delta}{G(-\theta)} [\theta G(-\theta) + \Phi(-\theta)] = 2\Delta \left[\imath + \frac{\Phi(-\theta)}{G(-\theta)} \right]$$

Now we shall calculate ξ'_θ.   We have

(A:121) $$\xi'_\theta = \operatorname*{Min}_r E_\theta(z + r \mid z + r \leq 0)$$

$$= - \operatorname*{Max}_r E_\theta(-z - r \mid -z - r \geq 0)$$

$$= -2\Delta \operatorname*{Max}_r E_\theta \left(-x - \frac{r}{2\Delta} \mid -x - \frac{r}{2\Delta} \geq 0 \right)$$

Let $t = -x + \theta$ and $t_0 = (r/2\Delta) + \theta$.   Then

(A:122) $$E_\theta \left(-x - \frac{r}{2\Delta} \mid -x - \frac{r}{2\Delta} \geq 0 \right) = E_\theta(t - t_0 \mid t - t_0 \geq 0)$$

$$= \frac{1}{G(t_0)} \int_{t_0}^{\infty} (t - t_0)\Phi(t) \, dt = \frac{\Phi(t_0)}{G(t_0)} - t_0$$

Since this is a monotonically decreasing function of t_0, we have

(A:123) $$\operatorname*{Max}_r E_\theta \left(-x - \frac{r}{2\Delta} \mid -x - \frac{r}{2\Delta} \geq 0 \right) = \frac{\Phi(\theta)}{G(\theta)} - \theta$$

From (A:121) and (A:123) we obtain

(A:124) $$\xi'_\theta = -2\Delta \left[\frac{\Phi(\theta)}{G(\theta)} - \theta \right]$$

Formulas (A:120) and (A:124) have been derived for the case when $\theta_0 = -\Delta$, $\theta_1 = \Delta$, and $\sigma = 1$. For general values θ_0, θ_1, and σ, the values of ξ_θ and ξ'_θ are given by

(A:125) $$\xi_\theta = \frac{1}{\sigma} (\theta_1 - \theta_0) \left[\bar{\theta} + \frac{\Phi(-\bar{\theta})}{G(-\bar{\theta})} \right]$$

and

(A:126) $$\xi'_\theta = -\frac{1}{\sigma} (\theta_1 - \theta_0) \left[\frac{\Phi(\bar{\theta})}{G(\bar{\theta})} - \bar{\theta} \right]$$

where

$$\bar{\theta} = \frac{1}{\sigma} \left(\theta - \frac{\theta_0 + \theta_1}{2} \right)$$

A.4 DERIVATION OF EXACT FORMULAS FOR THE OC AND ASN FUNCTIONS WHEN z CAN TAKE ONLY A FINITE NUMBER OF INTEGRAL MULTIPLES OF A CONSTANT

In this section we shall derive exact formulas for the OC and ASN functions when $z = \log \dfrac{f(x, \theta_1)}{f(x, \theta_0)}$ can take only a finite number of integral values of a positive constant d. This is a rather general result, since any distribution of z can be approximated arbitrarily closely by a discrete distribution of the above type if the constant d is chosen sufficiently small.

To obtain the exact OC and ASN functions, we shall first derive the exact probability distribution of the cumulative sum $Z_n = z_1 + \cdots + z_n$ at the termination of the sequential process. In what follows in this section the probability of any relation and the expected value of any random variable are determined under the assumption that θ is the true value of the parameter.[1] However, to simplify notation, we shall not put this in evidence in the formulas, i.e., we shall write P instead of P_θ and E instead of E_θ. Let g_1 and g_2 be two positive integers such that $P(z = -g_1 d)$ and $P(z = g_2 d)$ are positive and z can take only integral multiples of d which are $\geq -g_1 d$ and $\leq g_2 d$. Denote $P(z = id)$ by h_i. Then the moment-generating function of z is given by

$$(A{:}127) \qquad E(e^{zt}) = \sum_{i=-g_1}^{g_2} h_i e^{tid} = \phi(t) \qquad \text{(say)}$$

To obtain the roots of the equation $\phi(t) = 1$, we let $e^{td} = u$ and solve the equation:

$$(A{:}128) \qquad \sum_{i=-g_1}^{g_2} h_i u^i = 1$$

Let g denote $g_1 + g_2$ and let the g roots of (A:128) be $u_1, \cdots, u_g$, respectively. We shall assume that no two roots are equal, i.e., $u_i \neq u_j$ for $i \neq j$. Substituting u_i for e^{td} in the fundamental identity (A:16) we obtain

$$(A{:}129) \qquad E(u_i^{\frac{Z_n}{d}}) = 1 \qquad (i = 1, \cdots, g)$$

Let $[a]$ be the smallest integer $\geq \log A/d$, and $[b]$ the largest integer $\leq (\log B)/d$. Then Z_n/d can take only the values

$$(A{:}130)$$

$$([b] - g_1 + 1), ([b] - g_1 + 2), \cdots, [b], [a], ([a] + 1), \cdots, ([a] + g_2 - 1)$$

---

[1] If there are several unknown parameters, θ denotes the set of all parameters.

Denote the g different values in (A:130) by $c_1, \cdots, c_g$, respectively. Furthermore, denote $P(Z_n = c_i d)$ by ξ_i. Then equations (A:129) can be written as

$$(A:131) \qquad \sum_{j=1}^{g} \xi_j u_i{}^{c_j} = 1 \qquad (i = 1, \cdots, g)$$

Let Δ be the determinant value of the matrix $\|u_i{}^{c_j}\|$ $(i, j = 1, \cdots, g)$ and let Δ_j be the determinant we obtain from Δ by substituting 1 for the elements in the jth column. If $\Delta \neq 0$, it follows from (A:131) that $P(Z_n = c_j d) = \xi_j$ is given by

$$(A:132) \qquad \xi_j = \frac{\Delta_j}{\Delta}$$

Thus, the probability $L(\theta)$ that the process will terminate with $Z_n \leqq \log B$ is given by

$$(A:133) \qquad L(\theta) = \sum_j \frac{\Delta_j}{\Delta}$$

where the summation is to be taken over all values j for which $dc_j \leqq \log B$. Equation (A:133) is an exact equation of the OC function.

From the probability distribution of Z_n we can easily derive the expected value $E_\theta(n)$ of n. In fact, in Section A.3 it has been shown that

$$E_\theta(n) = \frac{E_\theta(Z_n)}{E_\theta(z)}$$

But

$$(A:134) \qquad E_\theta(Z_n) = \sum_{j=1}^{g} \frac{c_j \Delta_j d}{\Delta}$$

Hence

$$(A:135) \qquad E_\theta(n) = \frac{1}{E_\theta(z)} \sum_{j=1}^{g} \frac{c_j \Delta_j d}{\Delta}$$

is the exact equation of the ASN function.

The method of obtaining the probabilities $\xi_1, \cdots, \xi_g$, as described above, requires the computation of the roots of the polynomial equation (A:128). This is not necessary, however, if a method given by Girshick is used.[2] Girshick proceeds as follows. Multiplying $(\sum_i h_i u^i - 1)$ by u^{g_1} and $(\sum_j \xi_j u^{c_j} - 1)$ by $u^{g_1 - [b] - 1}$, we obtain two polynomials $f(u)$ and $F(u)$, where $f(u)$ is of degree $g_1 + g_2 = g$ and

---

[2] M. A. Girshick, "Contributions to the Theory of Sequential Analysis," *The Annals of Mathematical Statistics*, Vol. 17 (1946).

$F(u)$ of degree $g + [a] - [b] - 2$. According to (A:128) and (A:131), every root of $f(u)$ is also a root of $F(u)$. Hence

$$F(u) = f(u)f^*(u)$$

where $f^*(u)$ is a polynomial of degree $[a] - [b] - 2$, i.e.,

$$f^*(u) = k_0 + k_1 u + \cdots + k_{[a]-[b]-2}\, u^{[a]-[b]-2}$$

Putting the coefficient of any power of u in $F(u)$ equal to the coefficient of the same power of u in $f(u)f^*(u)$, we obtain a system of $g + [a] - [b] - 1$ linear equations in the $g + [a] - [b] - 1$ unknowns $\xi_1, \cdots, \xi_g, k_0, k_1, \cdots, k_{[a]-[b]-2}$, from which these unknowns can be determined. Thus, the probabilities $\xi_1, \cdots, \xi_g$ can be determined without solving the polynomial equation (A:128). This advantage is, however, bought for the price of an increased number of linear equations to be solved. If the roots of the polynomial equation (A:128) are computed, only g linear equations have to be solved for determining $\xi_1, \cdots, \xi_g$. If Girshick's method is used, no polynomial equation is to be solved, but the number of linear equations is increased to $g + [a] - [b] - 1$.

If $g_2 = 1$, the OC function $L(\theta)$ is a simple expression of the roots $u_1, \cdots, u_g$. In fact, $L(\theta) = P(Z_n \leqq \log B) = 1 - P(Z_n \geqq \log A) = 1 - \xi_g$. We have

$$\Delta = \begin{vmatrix} u_1^{[b]-g_1+1} & \cdots & u_1^{[b]} & u_1^{[a]} \\ \cdot & \cdots & \cdot & \cdot \\ \cdot & \cdots & \cdot & \cdot \\ u_g^{[b]-g_1+1} & \cdots & u_g^{[b]} & u_g^{[a]} \end{vmatrix}$$

and

$$\Delta_g = \begin{vmatrix} u_1^{[b]-g_1+1} & \cdots & u_1^{[b]} & 1 \\ \cdot & \cdots & \cdot & \cdot \\ \cdot & \cdots & \cdot & \cdot \\ u_g^{[b]-g_1+1} & \cdots & u_g^{[b]} & 1 \end{vmatrix}$$

The value of the ratio Δ_g/Δ is not changed if we multiply the ith row of Δ, as well as that of Δ_g, by $u_i^{g_1-[b]-1}$. Thus

$$\xi_g = \frac{\Delta_g}{\Delta} = \frac{\begin{vmatrix} 1 & u_1 & \cdots & u_1^{g_1-1} & u_1^{g_1-1-[b]} \\ \cdot & & \cdots & & \cdot \\ \cdot & & \cdots & & \cdot \\ 1 & u_g & \cdots & u_g^{g_1-1} & u_g^{g_1-1-[b]} \end{vmatrix}}{\begin{vmatrix} 1 & u_1 & \cdots & u_1^{g_1-1} & u_1^{g_1-1+[a]-[b]} \\ \cdot & & \cdots & & \cdot \\ \cdot & & \cdots & & \cdot \\ 1 & u_g & \cdots & u_g^{g_1-1} & u_g^{g_1-1+[a]-[b]} \end{vmatrix}}$$

The cofactor of each element in the last column is a Vandermonde determinant. Expanding the determinants in the numerator and denominator according to their last columns and dividing numerator and denominator by the Vandermonde determinant,

$$
\begin{vmatrix}
1 & u_1 & u_1{}^2 & \cdots & u_1{}^{g_1} \\
\multicolumn{5}{c}{\cdot\ \cdot\ \cdot\ \cdot\ \cdot\ \cdot\ \cdot\ \cdot\ \cdot} \\
\multicolumn{5}{c}{\cdot\ \cdot\ \cdot\ \cdot\ \cdot\ \cdot\ \cdot\ \cdot\ \cdot} \\
1 & u_g & u_g{}^2 & \cdots & u_g{}^{g_1}
\end{vmatrix}
\qquad (g_1 = g - 1)
$$

we obtain

$$
\xi_g = \frac{\Delta_g}{\Delta} = \frac{\displaystyle\sum_{i=1}^{g} \left[\frac{u_i{}^{g_1 - 1 - [b]}}{(u_i - 1)\displaystyle\prod_{j \neq i}(u_i - u_j)} \right]}{\displaystyle\sum_{i=1}^{g} \left[\frac{u_i{}^{g_1 - 1 + [a] - [b]}}{(u_i - 1)\displaystyle\prod_{j \neq i}(u_i - u_j)} \right]}
$$

We shall illustrate the derivation of the exact OC and ASN functions by a simple example. Let x be a random variable which can take only the values 0 and 1. Denote by H_i $(i = 0, 1)$ the hypothesis that the probability that $x = 1$ is equal to p_i $(i = 0, 1)$. Let

$$
p_0 = \frac{1 - e^{-2}}{e - e^{-2}} \quad \text{and} \quad p_1 = \frac{e - e^{-1}}{e - e^{-2}}
$$

Consider the sequential test for testing H_0 against H_1. We shall compute the probability that the process will terminate with the acceptance of H_0, and the expected number of trials required by the test, when the true probability that $x = 1$ is equal to $p = \frac{3}{4}$. In what follows in this section, all probability statements and expected values refer to the case when $p = \frac{3}{4}$.

First we compute $\phi(t) = E(e^{zt})$. Since z can take only the values

$$
\log \frac{p_1}{p_0} = \log e = 1 \quad \text{and} \quad \log \frac{1 - p_1}{1 - p_0} = \log e^{-2} = -2
$$

with probabilities $\frac{3}{4}$ and $\frac{4}{7}$, respectively, we have

$$
\phi(t) = \tfrac{3}{7}e^t + \tfrac{4}{7}e^{-2t}
$$

Letting $e^t = u$ and solving the equation

$$
\frac{3}{7}u + \frac{4}{7}\frac{1}{u^2} = 1
$$

we obtain the roots $u_1 = 1$, $u_2 = 2$, and $u_3 = -\tfrac{2}{3}$. The integers c_1, c_2, c_3 are given by

$$c_1 = \log B - 1, \quad c_2 = \log B, \quad c_3 = \log A$$

Hence

$$\Delta = \begin{vmatrix} 1 & 1 & 1 \\ 2^{\log B - 1} & 2^{\log B} & 2^{\log A} \\ (-\tfrac{2}{3})^{\log B - 1} & (-\tfrac{2}{3})^{\log B} & (-\tfrac{2}{3})^{\log A} \end{vmatrix}$$

$$\Delta_1 = \begin{vmatrix} 1 & 1 & 1 \\ 1 & 2^{\log B} & 2^{\log A} \\ 1 & (-\tfrac{2}{3})^{\log B} & (-\tfrac{2}{3})^{\log A} \end{vmatrix}$$

$$\Delta_2 = \begin{vmatrix} 1 & 1 & 1 \\ 2^{\log B - 1} & 1 & 2^{\log A} \\ (-\tfrac{2}{3})^{\log B - 1} & 1 & (-\tfrac{2}{3})^{\log A} \end{vmatrix}$$

$$\Delta_3 = \begin{vmatrix} 1 & 1 & 1 \\ 2^{\log B - 1} & 2^{\log B} & 1 \\ (-\tfrac{2}{3})^{\log B - 1} & (-\tfrac{2}{3})^{\log B} & 1 \end{vmatrix}$$

Then the probability that H_0 will be accepted is given by

$$L = \frac{\Delta_1 + \Delta_2}{\Delta}$$

The expected value of n is given by

$$\mathrm{E}(n) = \frac{1}{E(z)} \frac{c_1 \Delta_1 + c_2 \Delta_2 + c_3 \Delta_3}{\Delta}$$

$$= -\frac{7}{5} \frac{-(-\log B + 1)\Delta_1 + (\log B)\Delta_2 + (\log A)\Delta_3}{\Delta}$$

$$= \frac{7}{5} \frac{(-\log B + 1)\Delta_1 + (-\log B)\Delta_2 - (\log A)\Delta_3}{\Delta}$$

A.5 THE CHARACTERISTIC FUNCTION AND HIGHER MOMENTS OF n

A.5.1 Derivation of Approximate Formulas Neglecting the Excess of the Cumulative Sum over the Boundaries

Let $\bar{Z}_n$ be a random variable defined as follows: $\bar{Z}_n = \log A$ if $Z_n = z_1 + \cdots + z_n \geqq \log A$, and $\bar{Z}_n = \log B$ if $Z_n \leqq \log B$. Denote the difference $\bar{Z}_n - Z_n$ by ϵ. Then ϵ is a random variable.

In what follows in this section we shall neglect ϵ, i.e., we shall substitute 0 for ϵ. No error is committed by doing so in the special case when z can take only two values, d and $-d$, and the ratios $(\log A)/d$ and $(\log B)/d$ are integers, since in this case ϵ is exactly 0. Apart from this special case ϵ will not be identical with the constant 0. However, the smaller $|E(z)|$ and $E(z^2)$, the smaller the error we commit by neglecting ϵ. In fact, for arbitrarily small positive numbers δ_1 and δ_2 the inequality $P(|\epsilon| \leq \delta_1) \geq 1 - \delta_2$ will hold if $|E(z)|$ and $E(z^2)$ are sufficiently small. Thus, in the limiting case when $E(z)$ and $E(z^2)$ approach 0, the random variable ϵ reduces to the constant 0.

As in the preceding section, all probability statements and expected values will refer to the case in which θ is the true parameter point, without putting this in evidence in the formulas by using θ as a subscript to the operators P and E. Let $\phi(t)$ be the moment generating function of z, i.e.,

$$\phi(t) = E(e^{zt})$$

To derive an approximation to the characteristic function of n, we shall consider the equation

(A:136) $$- \log \phi(t) = \tau$$

where τ is a purely imaginary quantity. It will be assumed that z satisfies the conditions of lemma A.1. Then, according to lemma A.1, the equation $- \log \phi(t) = 0$ has exactly two real roots in t; they are $t = 0$ and $t = h$ ($h \neq 0$). Furthermore $\phi'(0)$ and $\phi'(h)$ both are unequal to 0. Hence, if $\phi(t)$ is not singular at $t = 0$ and $t = h$, equation (A:136) has two roots, $t_1(\tau)$ and $t_2(\tau)$, for sufficiently small values of $|\tau|$ such that $\lim_{\tau=0} t_1(\tau) = 0$ and $\lim_{\tau=0} t_2(\tau) = h$. Identity (A:16) can be written as

(A:137) $$LE^*\{e^{Z_n t}[\phi(t)]^{-n}\} + (1 - L)E^{**}\{e^{Z_n t}[\phi(t)]^{-n}\} = 1$$

where L denotes the probability that the test procedure leads to the acceptance of H_0, E^* stands for conditional expected value under the restriction that the process leads to the acceptance of H_0, E^{**} stands for conditional expected value under the restriction that the process leads to the rejection of H_0. Neglecting the excess of Z_n over the boundaries, we have $Z_n = \log B$ when the process leads to the acceptance of H_0, and $Z_n = \log A$ when the process leads to the rejection of H_0. Hence (A:137) can be written as

(A:138) $$LB^t E^*[\phi(t)]^{-n} + (1 - L)A^t E^{**}[\phi(t)]^{-n} = 1$$

This identity is valid for all values of t for which $|\phi(t)| \geqq 1$.[1] Letting $t = t_1(\tau)$ and $t = t_2(\tau)$, we obtain, from (A:138),

$$(A:139) \qquad LB^{t_1(\tau)}E^*(e^{\tau n}) + (1 - L)A^{t_1(\tau)}E^{**}(e^{\tau n}) = 1$$

and

$$(A:140) \qquad LB^{t_2(\tau)}E^*(e^{\tau n}) + (1 - L)A^{t_2(\tau)}E^{**}(e^{\tau n}) = 1$$

Solving these equations in $E^*(e^{\tau n})$ and $E^{**}(e^{\tau n})$, we obtain

$$(A:141) \qquad E^*(e^{\tau n}) = \frac{A^{t_2(\tau)} - A^{t_1(\tau)}}{L[B^{t_1(\tau)}A^{t_2(\tau)} - A^{t_1(\tau)}B^{t_2(\tau)}]}$$

and

$$(A:142) \qquad E^{**}(e^{\tau n}) = \frac{B^{t_1(\tau)} - B^{t_2(\tau)}}{(1 - L)[B^{t_1(\tau)}A^{t_2(\tau)} - A^{t_1(\tau)}B^{t_2(\tau)}]}$$

for all imaginary values τ.

The unconditional expected value $E(e^{\tau n})$ is clearly equal to

$$(A:143) \qquad E(e^{\tau n}) = LE^*(e^{\tau n}) + (1 - L)E^{**}(e^{\tau n})$$

Hence, the characteristic function of n is given by

$$(A:144) \qquad \psi(\tau) = E(e^{\tau n}) = \frac{A^{t_2(\tau)} - A^{t_1(\tau)} + B^{t_1(\tau)} - B^{t_2(\tau)}}{B^{t_1(\tau)}A^{t_2(\tau)} - A^{t_1(\tau)}B^{t_2(\tau)}}.$$

(for all imaginary τ).

By definition, the expected value $E(e^{\tau n})$ is the characteristic function of n, and (A:144) gives the desired approximation formula when the excess of Z_n over the boundaries can be neglected. Our derivations yield also approximation formulas for $\psi^*(\tau) = E^*(e^{\tau n})$ and $\psi^{**}(\tau) = E^{**}(e^{\tau n})$. The function $\psi^*(\tau)$ can be interpreted as the characteristic function of the conditional distribution of n when the process leads to the acceptance of H_0, and $\psi^{**}(\tau)$ can be interpreted as the characteristic function of the distribution of n in the subpopulation of samples leading to the rejection of H_0.

As an illustration we shall determine $\psi^*(\tau)$, $\psi^{**}(\tau)$, and $\psi(\tau)$ when z has a normal distribution. Denote by μ the mean of z and by σ the standard deviation of z. Then equation (A:136) can be written as

$$- \log \phi(t) = -\mu t - \frac{\sigma^2}{2}t^2 = \tau$$

[1] This follows from the considerations in Section A.2.2, since D' is the whole complex plane in our case.

Hence

(A:145)
$$t = \frac{-\mu \pm \sqrt{\mu^2 - 2\sigma^2\tau}}{\sigma^2}$$

Thus

(A:146)
$$t_1(\tau) = -\frac{\mu}{\sigma^2} + \frac{1}{\sigma^2}\sqrt{\mu^2 - 2\sigma^2\tau}$$

and

(A:147)
$$t_2(\tau) = -\frac{\mu}{\sigma^2} - \frac{1}{\sigma^2}\sqrt{\mu^2 - 2\sigma^2\tau}$$

where the sign of $\sqrt{}$ is determined so that the real part of $\sqrt{\mu^2 - 2\sigma^2\tau}$ is positive. Substituting these values for $t_1(\tau)$ and $t_2(\tau)$ in (A:141), (A:142), and (A:144), we obtain $\psi^*(\tau)$, $\psi^{**}(\tau)$, and $\psi(\tau)$ in the case when z is normally distributed. According to formula (3:43), an approximation to L is given by

(A:148)
$$L \sim \frac{A^h - 1}{A^h - B^h}$$

When z is normally distributed we have

(A:149)
$$h = \frac{-2\mu}{\sigma^2}$$

It is of interest to consider the following two limiting cases: (1) $B = 0$ and A is a finite positive value; (2) B is a finite positive value and $A = +\infty$. It can be shown that $E(n)$ will be finite in case (1) only if $E(z) > 0$. Similarly, $E(n)$ will be finite in case (2) only if $E(z) < 0$. Thus, in case (1) we shall assume that $E(z) > 0$, and in case (2) we shall assume that $E(z) < 0$. To obtain the characteristic function $\psi(\tau)$ of n in case (1), we have to determine the limiting value of the right-hand member of (A:144) when $B \to 0$. For this purpose we shall first derive the limiting value of $B^{t_2(\tau)}/B^{t_1(\tau)} = B^{t_2(\tau)-t_1(\tau)}$ when $B \to 0$. Since in case (1) $E(z)$ is assumed to be > 0, the quantity $h = \lim_{\tau=0} t_2(\tau)$ must be negative, as has been shown in Section A.2.1. Hence, for small τ the real part of $t_2(\tau)$ is negative. On the other hand, the real part of $t_1(\tau)$ approaches 0 as $\tau \to 0$. Thus, for small τ the real part of $t_2(\tau) - t_1(\tau)$ is negative, and, therefore,

(A:150)
$$\lim_{B=0} \left| B^{t_2(\tau)-t_1(\tau)} \right| = +\infty$$

From (A:150) and from the relation $\lim_{B=0} \left| B^{t_2(\tau)} \right| = \infty$, it follows that with $B \to 0$ the right-hand member of (A:144) converges to

(A:151)
$$A^{-t_1(\tau)}$$

Thus, if $E(z) > 0$ the characteristic function of n in case (1) is given by (A:151). When z is normally distributed, $t_1(\tau)$ is given by (A:146). Hence, for normally distributed z with $\mu > 0$ the characteristic function of n in case (1) is given by

$$(A{:}152) \qquad A^{\frac{\mu}{\sigma^2} - \frac{1}{\sigma^2}\sqrt{\mu^2 - 2\sigma^2\tau}}$$

In case (2) we have assumed that $E(z) < 0$. Hence $t_2(\tau)$ and $t_2(\tau) - t_1(\tau)$ will have a positive real part for small τ. Thus,

$$(A{:}153) \qquad \lim_{A=\infty} \left| A^{t_2(\tau)} \right| = \lim_{A=\infty} \left| A^{t_2(\tau)-t_1(\tau)} \right| = +\infty$$

From (A:153) it follows that the limiting value of the right-hand member of (A:144) when $A \to \infty$ is given by

$$(A{:}154) \qquad B^{-t_1(\tau)}$$

Thus, if $E(z) < 0$, the characteristic function of n in case (2) is given by (A:154).

The moments of n can be obtained by differentiating the characteristic function of n. For any positive integer r the rth moment of n is given by

$$(A{:}155) \qquad E(n^r) = \frac{d^r}{d\tau^r}\psi(\tau).$$

We can also obtain the conditional moments of n in the subpopulation of samples for which $Z_n \le \log B$, as well as in the subpopulation of samples for which $Z_n \ge \log A$. Let $E^*(n^r)$ denote the conditional expected value of n^r in the subpopulation $Z_n \le \log B$, and let $E^{**}(n^r)$ denote the expected value of n^r in the subpopulation $Z_n \ge \log A$. Then we have

$$E^*(n^r) = \frac{d^r}{d\tau^r}\psi^*(\tau) \quad \text{and} \quad E^{**}(n^r) = \frac{d^r}{d\tau^r}\psi^{**}(\tau)$$

where $\psi^*(\tau)$ and $\psi^{**}(\tau)$ are the conditional characteristic functions given in (A:141) and (A:142).

It may be of interest to note that $\frac{d^r}{d\tau^r}\psi^*(\tau)$, $\frac{d^r}{d\tau^r}\psi^{**}(\tau)$, and, therefore, also $E(n^r) = \frac{d^r}{d\tau^r}\psi(\tau)$ can be obtained from identity (A:138) directly by successive differentiation. In fact, (A:138) can be written as

$$(A{:}156) \quad LB^t\psi^*[-\log\phi(t)] + (1-L)A^t\psi^{**}[-\log\phi(t)] = 1$$

Taking the first r derivatives of (A:156) with respect to t at $t = 0$ and $t = h$, we obtain a system of $2r$ linear equations in the $2r$ unknowns $\dfrac{d^j}{d\tau^j}\psi^*(\tau)\Big|_{\tau=0}$ and $\dfrac{d^j}{d\tau^j}\psi^{**}(\tau)\Big|_{\tau=0}$ $(j = 1, \cdots, r)$ from which these unknowns can be determined. For example $\dfrac{d\psi^*(\tau)}{d\tau}\Big|_{\tau=0}$ and $\dfrac{d\psi^{**}(\tau)}{d\tau}\Big|_{\tau=0}$ can be determined as follows. Taking the first derivative of (A:156) with respect to t we obtain

$$(A:157) \quad L(\log B)B^t\psi^*(\tau) - LB^t\frac{\phi'(t)}{\phi(t)}\frac{d\psi^*(\tau)}{d\tau} +$$
$$(1 - L)(\log A)A^t\psi^{**}(\tau) - (1 - L)A^t\frac{\phi'(t)}{\phi(t)}\frac{d\psi^{**}(\tau)}{d\tau} = 0$$
$$[\tau = -\log\phi(t)]$$

Letting $t = 0$ and $t = h$ we obtain the equations

$$(A:158) \quad L\log B - L\frac{\phi'(0)}{\phi(0)}\frac{d\psi^*(\tau)}{d\tau}\Big|_{\tau=0} +$$
$$(1 - L)\log A - (1 - L)\frac{\phi'(0)}{\phi(0)}\frac{d\psi^{**}(\tau)}{d\tau}\Big|_{\tau=0} = 0$$

and

$$(A:159) \quad L(\log B)B^h - LB^h\frac{\phi'(h)}{\phi(h)}\frac{d\psi^*(\tau)}{d\tau}\Big|_{\tau=0} + (1 - L)(\log A)A^h -$$
$$(1 - L)A^h\frac{\phi'(h)}{\phi(h)}\frac{d\psi^{**}(\tau)}{d\tau}\Big|_{\tau=0} = 0$$

from which $\dfrac{d\psi^*(\tau)}{d\tau}\Big|_{\tau=0}$ and $\dfrac{d\psi^{**}(\tau)}{d\tau}\Big|_{\tau=0}$ can be determined.

A.5.2 Derivation of Exact Formulas When z Can Take Only a Finite Number of Integral Multiples of a Constant

We shall use here the notation defined in Section A.4 without any further explanation. Let $\psi_i(\tau)$ denote the characteristic function of the conditional distribution of n in the subpopulation of samples for which $Z_n = c_i d$ $(i = 1, \cdots, g)$. The equation in t

$$(A:160) \qquad \phi(t) = e^{-\tau}$$

has g roots $t_1(\tau), \cdots, t_g(\tau)$ such that

$$(A:161) \qquad \lim_{\tau=0} e^{t_i(\tau)d} = u_i \qquad (i = 1, \cdots, g)$$

The fundamental identity (A:16) can be written as

(A:162) $$\sum_{j=1}^{g} \xi_j e^{c_j t d} \psi_j [- \log \phi(t)] = 1$$

Substituting $t_i(\tau)$ for t in (A:162), we obtain

(A:163) $$\sum_{j=1}^{g} \xi_j e^{c_j t_i(\tau) d} \psi_j(\tau) = 1 \qquad (i = 1, \cdots, g)$$

These equations are linear in the unknowns $\psi_1(\tau), \cdots, \psi_g(\tau)$, and the determinant of these equations is given by

(A:164) $$\delta(\tau) = \begin{vmatrix} \xi_1 e^{c_1 t_1(\tau) d} & \cdots & \xi_g e^{c_g t_1(\tau) d} \\ \xi_1 e^{c_1 t_2(\tau) d} & \cdots & \xi_g e^{c_g t_2(\tau) d} \\ \cdots & \cdots & \cdots \\ \xi_1 e^{c_1 t_g(\tau) d} & \cdots & \xi_g e^{c_g t_g(\tau) d} \end{vmatrix}$$

Obviously, $\delta(0) = \xi_1 \xi_2 \cdots \xi_g \Delta$. Hence, if $\xi_i \neq 0$ $(i = 1, \cdots, g)$ and $\Delta \neq 0$, then $\delta(0) \neq 0$, and consequently $\delta(\tau) \neq 0$ for any τ with sufficiently small absolute value. Thus, $\psi_1(\tau), \cdots, \psi_g(\tau)$ can be obtained by solving the linear equations (A:163).[1] The characteristic function $\psi(\tau)$ of the unconditional distribution of n is given by

(A:165) $$\psi(\tau) = \sum_{i=1}^{g} \xi_i \psi_i(\tau)$$

For any positive integer r, the exact rth moment of n, i.e., $E(n^r)$, is given by the rth derivative of $\psi(\tau)$ with respect to τ at $\tau = 0$.

A.6 APPROXIMATE DISTRIBUTION OF n WHEN z IS NORMALLY DISTRIBUTED

A.6.1 The Case When $B = 0$ and A Is Finite

In this case we have assumed that $E(z) = \mu > 0$. Then the approximate characteristic function of n, if the excess of Z_n over the boundaries is neglected, is given by (A:152). Let

(A:166) $$m = \frac{\mu^2}{2\sigma^2} n$$

---

[1] This method of determining $\psi_1(\tau), \cdots, \psi_g(\tau)$ requires the computation of the roots of equation (A:160). This can be avoided, as Girshick has shown in his paper mentioned in Section A.4, if a device is used similar to that applied by him for determining $\xi_1, \cdots, \xi_g$ (see Section A.4).

Then the characteristic function of m is given by

(A:167) $$\bar{\psi}(t) = e^{c[1-\sqrt{1-t}]}$$

where

(A:168) $$c = \frac{a\mu}{\sigma^2} > 0$$

and

(A:169) $$a = \log A$$

The sign of the square root in (A:167) is determined so that the real part of $\sqrt{1-t}$ is positive. The distribution of m is given by

(A:170) $$\frac{1}{2\pi i}\int_{-i\infty}^{i\infty} e^{c(1-\sqrt{1-t})-mt}\, dt$$

Let

(A:171) $$G(c, m) = \frac{1}{2\pi i}\int_{-i\infty}^{i\infty} e^{-c\sqrt{1-t}-mt}\, dt$$

and

(A:172) $$H(c, m) = \frac{1}{2\pi i}\int_{-i\infty}^{i\infty} \frac{1}{\sqrt{1-t}} e^{-c\sqrt{1-t}-mt}\, dt$$

Since

(A:173) $$\frac{1}{2\pi i}\frac{d}{dt} e^{-c\sqrt{1-t}-mt} = \frac{1}{2\pi i}\left(\frac{c}{2\sqrt{1-t}} - m\right) e^{-c\sqrt{1-t}-mt}$$

we have

(A:174) $$\frac{c}{2} H(c, m) - mG(c, m) = \frac{1}{2\pi i}\left[e^{-c\sqrt{1-t}-mt}\right]_{-i\infty}^{i\infty} = 0$$

From (A:171) and (A:172) we obtain

(A:175) $$\frac{\partial H(c, m)}{\partial c} + G(c, m) = 0$$

From (A:174) and (A:175) it follows that

(A:176) $$\frac{c}{2} H(c, m) + m\frac{\partial H(c, m)}{\partial c} = 0$$

Hence

(A:177) $$\log H(c, m) = -\frac{c^2}{4m} + \log \lambda(m)$$

where $\lambda(m)$ is some function of m only. Thus

(A:178) $$H(c, m) = \lambda(m)e^{-\frac{c^2}{4m}}$$

Now we shall determine $\lambda(m)$. We have

$$(\text{A:179}) \qquad \lambda(m) = H(0, m) = \frac{1}{2\pi i} \int_{-i\infty}^{i\infty} \frac{1}{\sqrt{1-t}} e^{-mt} \, dt$$

Since $(1-t)^{-\frac{1}{2}}$ is the characteristic function of $\frac{1}{2}\chi^2$ where χ^2 has the χ^2-distribution with one degree of freedom, the right-hand side of (A:179) is equal to

$$\frac{1}{\Gamma(\frac{1}{2})\sqrt{m}} e^{-m}$$

Hence

$$(\text{A:180}) \qquad \lambda(m) = \frac{1}{\Gamma(\frac{1}{2})\sqrt{m}} e^{-m}$$

From (A:178) and (A:179) we obtain

$$(\text{A:181}) \qquad H(c, m) = \frac{1}{\Gamma(\frac{1}{2})\sqrt{m}} e^{-\frac{c^2}{4m} - m}$$

From (A:174) and (A:181) we obtain

$$(\text{A:182}) \qquad G(c, m) = \frac{c}{2\Gamma(\frac{1}{2})m^{3/2}} e^{-\frac{c^2}{4m} - m}$$

Hence the distribution of m is given by

$$(\text{A:183}) \quad F(m) \, dm = \frac{c}{2\Gamma(\frac{1}{2})m^{3/2}} e^{-\frac{c^2}{4m} - m + c} \, dm \qquad (0 \leqq m < \infty)$$

Let $m = (c/2)m^*$. Then the distribution of m^* is given by

$$(\text{A:184}) \quad D(m^*) \, dm^* = \frac{\dfrac{c^2}{2}}{2\Gamma\left(\dfrac{1}{2}\right)\left(\dfrac{c}{2}\right)^{3/2}(m^*)^{3/2}} e^{-\frac{c}{2}\left(\frac{1}{m^*} + m^* - 2\right)} \, dm^*$$

$$= \frac{\sqrt{c}}{\sqrt{2\pi}(m^*)^{3/2}} e^{-\frac{c}{2}\left(\frac{1}{m^*} + m^* - 2\right)} \, dm^*$$

The function $(1/m^*) + m^* - 2$ is non-negative and is equal to 0 only when $m^* = 1$. If c is large, then $D(m^*)$ is exceedingly small for values of m^* not close to 1. Expanding $(1/m^*) + m^* - 2$ in a Taylor series around $m^* = 1$, we obtain

$$(\text{A:185}) \quad \frac{1}{m^*} + m^* - 2 = (m^* - 1)^2 + \text{higher order terms}$$

Hence for large c

(A:186) $$D(m^*) \, dm^* \sim \frac{\sqrt{c}}{\sqrt{2\pi}} \, e^{-\left(\frac{c}{2}\right)(m^*-1)^2} \, dm^*$$

i.e., if c is large m^* is nearly normally distributed with mean equal to 1 and standard deviation $1/\sqrt{c}$.

A.6.2   The Case When $B > 0$ and $A = \infty$

In this case we have assumed that $E(z) = \mu < 0$. It can easily be shown that the distribution of $m = (\mu^2/2\sigma^2)n$ is now given by the expression we obtain from (A:183) if we substitute $(\mu/\sigma^2) \log B$ for c.

A.6.3   The Case When $B > 0$ and A Is Finite

In this case the approximate characteristic function of n, if the excess of Z_n over the boundaries is neglected, is given by (A:144) where $t_1(\tau)$ and $t_2(\tau)$ are equal to the right-hand members of (A:146) and (A:147), respectively.   Let

$$m = \frac{\mu^2}{2\sigma^2} n \quad \text{and} \quad d = -\frac{\mu}{\sigma^2}$$

Then the characteristic function of m is given by

(A:187) $$\psi(t) = \frac{A^{h_1} + B^{h_2} - A^{h_2} - B^{h_1}}{A^{h_1} B^{h_2} - A^{h_2} B^{h_1}}$$

where

(A:188) $$h_1 = d(1 - \sqrt{1-t}), \quad h_2 = d(1 + \sqrt{1-t})$$

and t is an imaginary variable. Letting $A^d = \bar{A}$, $B^d = \bar{B}$, $da = \bar{a}$, and $db = \bar{b}$, the characteristic function of m can be written as

(A:189)

$$\psi(t) = \frac{\bar{A}(e^{-\bar{a}\sqrt{1-t}} - e^{\bar{a}\sqrt{1-t}}) + \bar{B}(e^{\bar{b}\sqrt{1-t}} - e^{-\bar{b}\sqrt{1-t}})}{\bar{A}\bar{B}(e^{(\bar{b}-\bar{a})\sqrt{1-t}} - e^{(\bar{a}-\bar{b})\sqrt{1-t}})}$$

$$= \frac{\bar{A}(e^{-\bar{b}\sqrt{1-t}} - e^{(2\bar{a}-\bar{b})\sqrt{1-t}}) + \bar{B}(e^{\bar{a}\sqrt{1-t}} - e^{(\bar{a}-2\bar{b})\sqrt{1-t}})}{\bar{A}\bar{B}(1 - e^{2(\bar{a}-\bar{b})\sqrt{1-t}})}$$

It will be sufficient to consider only the case when $\mu > 0$, since the case when $\mu < 0$ can be treated in a similar way. Then $\bar{a} < 0$ and

$\bar{b} > 0$. Since the real part of $+\sqrt{1-t}$ is greater than or equal to 1, we have

(A:190) $$\left| e^{2(\bar{a}-\bar{b})\sqrt{1-t}} \right| < 1$$

for any imaginary value of t. Let

(A:191) $$T = e^{2(\bar{a}-\bar{b})\sqrt{1-t}}$$

Then

(A:192) $$\frac{1}{1-T} = \sum_{j=0}^{\infty} T^j$$

From (A:189) and (A:192) it follows that $\bar{\psi}(t)$ can be written in the form of an infinite series:

(A:193) $$\bar{\psi}(t) = \sum_{i=1}^{\infty} r_i e^{-\lambda_i \sqrt{1-t}}$$

where λ_i and r_i are constants and $\lambda_i > 0$. Each term of this series is a characteristic function of the form given in (A:167) except for a proportionality factor. Let $F_i(m)$ be the distribution of m corresponding to the characteristic function $e^{\lambda_i - \lambda_i \sqrt{1-t}}$. Then $F_i(m)$ can be obtained from (A:183) by substituting λ_i for c. Since we may integrate the right-hand member of (A:193) term by term, the distribution of m is given by

(A:194) $$F(m)\, dm = \left[\sum_{i=1}^{\infty} \frac{r_i}{e^{\lambda_i}} F_i(m) \right] dm$$

A.6.4 Some Remarks

Since m is a discrete variable, it may seem paradoxical that we obtained a probability density function for m. However, the explanation lies in the fact that we neglected $\epsilon = \bar{Z}_n - Z_n$ and this quantity is 0 only in the limiting case when μ and σ approach 0.

If $|\mu|$ and σ are sufficiently small as compared with $\log A$ and $|\log B|$, the distribution of m given in (A:194) will be a good approximation to the exact distribution of m, even if z is not normally distributed. The reason for this can be indicated as follows. Let

(A:195) $$z_i^* = \sum_{j=(i-1)r+1}^{ir} z_j \qquad (i = 1, 2, \cdots, \text{ad inf.})$$

where r is a given positive integer. Since the variates z_j are independently distributed, each having the same distribution, under some weak conditions the variates z_i^* $(i = 1, 2, \cdots, \text{ad inf.})$ will be nearly normally distributed for large r. Hence, considering the cumulative sums $Z_i{}^* = z_1{}^* + z_2{}^* + \cdots + z_i{}^*$ $(i = 1, 2, \cdots, \text{ad inf.})$, the distribution given in (A:194) is applicable with good approximation, provided that $r \mid \mu \mid$ and $\sqrt{r}\sigma$ are small compared with $\log A$ and $\mid \log B \mid$ so that the difference $\epsilon^* = \bar{Z}_n{}^* - Z_n{}^*$ can be neglected.

It would be desirable to derive limits for the error in the cumulative distribution of m caused by neglecting $\bar{Z}_n - Z_n$. No such limits have yet been obtained.

A.7 EFFICIENCY OF THE SEQUENTIAL PROBABILITY RATIO TEST

Let S be any sequential test for testing H_0 against H_1 such that the probability of an error of the first kind is α, and the probability of an error of the second kind is β, and the probability that the test procedure will eventually terminate is 1. Let S' be the sequential probability ratio test whose strength is equal to that of S. We shall prove that the sequential probability ratio test is an optimum test, i.e., that $E_i(n \mid S) \geqq E_i(n \mid S')$ $(i = 0, 1)$, if for S' the excess of Z_n over $\log A$ and $\log B$ can be neglected.[1] This excess is exactly 0 if z can take only the values d and $-d$ and if $\log A$ and $\log B$ are integral multiples of d. In any other case the excess will not be identically 0. However, if $\mid E(z) \mid$ and the standard deviation σ_z of z are sufficiently small, the excess of Z_n over $\log A$ and $\log B$ is negligible.

For any random variable u, we shall denote by $E_i{}^*(u \mid S)$ the conditional expected value of u under the hypothesis H_i $(i = 0, 1)$ and under the restriction that H_0 is accepted. Similarly, let $E_i{}^{**}(u \mid S)$ be the conditional expected value of u under the hypothesis H_i $(i = 0, 1)$ and under the restriction that H_1 is accepted. In the notations for these expected values, the symbol S stands for the sequential test used. Let $Q_i(S)$ denote the totality of all samples for which the test S leads to the acceptance of H_i. Then we have

$$(\text{A:196}) \qquad E_0{}^* \left(\frac{p_{1n}}{p_{0n}} \,\middle|\, S \right) = \frac{P_1[Q_0(S)]}{P_0[Q_0(S)]} = \frac{\beta}{1 - \alpha}$$

$$(\text{A:197}) \qquad E_0{}^{**} \left(\frac{p_{1n}}{p_{0n}} \,\middle|\, S \right) = \frac{P_1[Q_1(S)]}{P_0[Q_1(S)]} = \frac{1 - \beta}{\alpha}$$

$$(\text{A:198}) \qquad E_1{}^* \left(\frac{p_{0n}}{p_{1n}} \,\middle|\, S \right) = \frac{P_0[Q_0(S)]}{P_1[Q_0(S)]} = \frac{1 - \alpha}{\beta}$$

and

$$(\text{A:199}) \qquad E_1{}^{**} \left(\frac{p_{0n}}{p_{1n}} \,\middle|\, S \right) = \frac{P_0[Q_1(S)]}{P_1[Q_1(S)]} = \frac{\alpha}{1 - \beta}$$

---

[1] $E_i(n \mid S)$ denotes the expected value of n when H_i is true $(\theta = \theta_i)$ and the sequential test S is used.

To prove the optimum property of the sequential probability ratio test, we shall first derive two lemmas.

Lemma A.2.   For any random variable u the inequality

(A:200) $$e^{E(u)} \leqq E(e^u)$$

holds.

Proof.   Inequality (A:200) can be written as

(A:201) $$1 \leqq E(e^{u'})$$

where $u' = u - E(u)$.   Lemma A.2 is proved if we show that (A:201) holds for any random variable u' with zero mean.   Expanding $e^{u'}$ in a Taylor series around $u' = 0$, we obtain

(A:202) $$e^{u'} = 1 + u' + \tfrac{1}{2}u'^2 e^{\xi(u')}$$

where $\xi(u')$ lies between 0 and u'.   Hence

(A:203) $$E(e^{u'}) = 1 + \tfrac{1}{2}E[u'^2 e^{\xi(u')}] \geqq 1$$

and lemma A.2 is proved.

Lemma A.3.   Let S be a sequential test such that there exists a finite integer N with the property that the number n of observations required for the test is $\leqq N$.   Then [2]

(A:204) $$E_i(n \mid S) = \frac{E_i \left(\log \dfrac{p_{1n}}{p_{0n}} \, \middle| \, S \right)}{E_i(z)} \qquad (i = 0, 1)$$

The proof is omitted, since it is essentially the same as that of equation (A:69) for the sequential probability ratio test.

On the basis of lemmas A.2 and A.3 we shall be able to derive the following theorem.

Theorem: *Let S be any sequential test for which the probability of an error of the first kind is α, the probability of an error of the second kind is β, and the probability that the test procedure will eventually terminate is equal to 1.   Then*

(A:205) $$E_0(n \mid S) \geqq \frac{1}{E_0(z)} \left[(1 - \alpha) \log \frac{\beta}{1 - \alpha} + \alpha \log \frac{1 - \beta}{\alpha} \right]$$

and

(A:206) $$E_1(n \mid S) \geqq \frac{1}{E_1(z)} \left[\beta \log \frac{\beta}{1 - \alpha} + (1 - \beta) \log \frac{1 - \beta}{\alpha} \right]$$

[2] The validity of (A:204) has been established under very general conditions even when the probability that $n > N$ is positive for any N. See the author's article, "Some Generalizations of the Theory of Cumulative Sums," *The Annals of Mathematical Statistics*, Vol. 16 (1945), and D. Blackwell, "On an Equation of Wald," *The Annals of Mathematical Statistics*, Vol. 17 (1946).

Proof. First we shall prove the theorem in the case when there exists a finite integer N such that n never exceeds N. According to lemma A.3 we have

$$(A{:}207) \quad E_0(n \mid S) = \frac{1}{E_0(z)} E_0 \left(\log \frac{p_{1n}}{p_{0n}} \mid S \right)$$

$$= \frac{1}{E_0(z)} \left[(1 - \alpha) E_0{}^* \left(\log \frac{p_{1n}}{p_{0n}} \mid S \right) + \alpha E_0{}^{**} \left(\log \frac{p_{1n}}{p_{0n}} \mid S \right) \right]$$

and

$$(A{:}208) \quad E_1(n \mid S) = \frac{1}{E_1(z)} E_1 \left(\log \frac{p_{1n}}{p_{0n}} \mid S \right)$$

$$= \frac{1}{E_1(z)} \left[\beta E_1{}^* \left(\log \frac{p_{1n}}{p_{0n}} \mid S \right) + (1 - \beta) E_1{}^{**} \left(\log \frac{p_{1n}}{p_{0n}} \mid S \right) \right]$$

From equations (A:196) through (A:199) and lemma A.2 we obtain the inequalities

$$(A{:}209) \quad E_0{}^* \left(\log \frac{p_{1n}}{p_{0n}} \mid S \right) \leq \log \frac{\beta}{1 - \alpha}$$

$$(A{:}210) \quad E_0{}^{**} \left(\log \frac{p_{1n}}{p_{0n}} \mid S \right) \leq \log \frac{1 - \beta}{\alpha}$$

$$(A{:}211) \quad E_1{}^* \left(\log \frac{p_{0n}}{p_{1n}} \mid S \right) = -E_1{}^* \left(\log \frac{p_{1n}}{p_{0n}} \mid S \right) \leq \log \frac{1 - \alpha}{\beta}$$

and

$$(A{:}212) \quad E_1{}^{**} \left(\log \frac{p_{0n}}{p_{1n}} \mid S \right) = -E_1{}^{**} \left(\log \frac{p_{1n}}{p_{0n}} \mid S \right) \leq \log \frac{\alpha}{1 - \beta}$$

Since $E_0(z) < 0$, (A:205) follows from (A:207), (A:209), and (A:210). Similarly, since $E_1(z) > 0$, (A:206) follows from (A:208), (A:211), and (A:212). This proves the theorem when a finite integer N exists such that $n \leq N$.

To prove the theorem for any sequential test S of strength (α, β), let S_N be the sequential test we obtain by truncating S at the Nth observation if no decision is reached before the Nth observation. Let (α_N, β_N) be the strength of S_N. Then we have

$$(A{:}213) \quad E_0(n \mid S) \geq E_0(n \mid S_N)$$

$$\geq \frac{1}{E_0(z)} \left[(1 - \alpha_N) \log \frac{\beta_N}{1 - \alpha_N} + \alpha_N \log \frac{1 - \beta_N}{\alpha_N} \right]$$

and

(A:214) $\quad E_1(n \mid S) \geq E_1(n \mid S_N)$

$$\geq \frac{1}{E_1(z)} \left[\beta_N \log \frac{\beta_N}{1 - \alpha_N} + (1 - \beta_N) \log \frac{1 - \beta_N}{\alpha_N} \right]$$

Since $\lim_{N=\infty} \alpha_N = \alpha$ and $\lim_{N=\infty} \beta_N = \beta$, inequalities (A:205) and (A:206) follow from (A:213) and (A:214). Hence the proof of the theorem is completed.

If for the sequential probability ratio test S' the excess of the cumulative sum Z_n over the boundaries $\log A$ and $\log B$ is 0, $E_0(n \mid S')$ is exactly equal to the right-hand member of (A:205) and $E_1(n \mid S')$ is exactly equal to the right-hand member of (A:206). Hence, in this case, S' is exactly an optimum test. If both $\mid E(z) \mid$ and σ_z are small, the expected value of the excess over the boundaries will also be small and, therefore, $E_0(n \mid S')$ and $E_1(n \mid S')$ will be only slightly larger than the right-hand members of (A:205) and (A:206), respectively. Thus, in such a case, the sequential probability ratio test is, if not exactly, very nearly an optimum test.[3]

If θ_1 approaches θ_0, then the ratios of the upper limits of $E_0(n \mid S')$ and $E_1(n \mid S')$, as implied by (A:77) and (A:78), to the right-hand members of (A:205) and (A:206), respectively, converge to 1. Thus, the efficiency of the sequential probability ratio test, if not exactly 1, converges to 1 when $\theta_1 \to \theta_0$.[4] The upper bounds for $E_0(n \mid S')$ and $E_1(n \mid S')$ given in (A:77) and (A:78) determine lower bounds for the efficiency of the sequential probability ratio test S'.

A.8 DETERMINATION OF AN OPTIMUM WEIGHT FUNCTION $w(\theta)$ IN SOME SPECIAL CASES OF TESTING SIMPLE HYPOTHESES WITH NO RESTRICTIONS ON THE POSSIBLE ALTERNATIVE VALUES OF THE PARAMETERS

A.8.1 A Class of Cases for Which an Optimum Weight Function $w(\theta)$ Can Be Determined by a Simple Procedure

Let $(\theta_1, \cdots, \theta_k) = (\theta_1{}^0, \cdots, \theta_k{}^0)$ be the simple hypothesis H_0 to be tested and denote the distribution of x by $f(x, \theta_1, \cdots, \theta_k)$. Assume the boundary of the zone ω_r of preference for rejection is a surface in the parameter space and denote it by S_r. Assume, further, that it is

---

[3] The author conjectures that the sequential probability ratio test is exactly an optimum test even if the excess of Z_n over the boundaries is not 0. However, he did not succeed in proving this.

[4] For the definition of the efficiency of a sequential test see Section 2.4.1.

possible to find a non-negative function $v(\theta)$ of the parameter θ such that the surface integral [1]

$$(A:215) \qquad\qquad \int_{S_r} v(\theta) \, dS = 1$$

and the sequential probability ratio test based on the ratio

$$(A:216) \qquad \frac{p_{1n}}{p_{0n}} = \frac{\int_{S_r} f(x_1, \theta_1, \cdots, \theta_k) \cdots f(x_n, \theta_1, \cdots, \theta_k) v(\theta) \, dS}{f(x_1, \theta_1^0, \cdots, \theta_k^0) \cdots f(x_n, \theta_1^0, \cdots, \theta_k^0)}$$

satisfies the following two conditions (for any values A and B): (1) The probability $\beta(\theta)$ of committing an error of the second kind (of accepting H_0 when θ is true) is constant over the surface S_r; (2) for any point θ in the interior of ω_r, the value of $\beta(\theta)$ does not exceed the constant value of $\beta(\theta)$ on the surface S_r.

We shall now show that $v(\theta)$ may be regarded as an optimum weight function in the sense defined in Section 4.1.3, and the probability ratio test based on the ratio (A:216) provides a solution to our problem. In fact, the weight function $v(\theta)$ over the surface S_r can be considered a limiting case of a weight function $w(\theta)$ which takes the value 0 for any θ in the interior of ω_r whose distance from the boundary exceeds some positive Δ, with Δ approaching 0 in the limit. It follows from conditions (1) and (2) that for the weight function $v(\theta)$ the maximum of $\beta(\theta)$ in ω_r is equal to the weighted integral of $\beta(\theta)$, i.e., to $\int_{S_r} \beta(\theta) v(\theta) \, dS$. Consider now any other weight function $w^*(\theta)$ and denote the resulting probability of an error of the second kind by $\beta^*(\theta)$ when $w^*(\theta)$ is used instead of $v(\theta)$. It has been shown in Section 4.1.3 that the following relations hold with sufficient approximation for practical purposes:

$$(A:217) \qquad \int_{\omega_r} w^*(\theta)\beta^*(\theta) \, d\theta = \int_{S_r} v(\theta)\beta(\theta) \, dS = \frac{B(A-1)}{A-B}$$

Hence the maximum of $\beta^*(\theta)$ in ω_r is $\geqq B(A-1)/(A-B)$. The optimum property of the weight function $v(\theta)$ follows then from the fact that the maximum of $v(\theta)$ is equal to $B(A-1)/(A-B)$.

In several important statistical problems one can easily find a weight function $v(\theta)$ such that conditions (1) and (2) are fulfilled. We shall show, for example, that such a weight function $v(\theta)$ can easily be determined for testing the means of normally distributed variables with

---

[1] dS denotes the infinitesimal surface element.

known variances. After the weight function $v(\theta)$ has been found, for practical purposes we may let $A = (1 - \beta)/\alpha$ and $B = \beta/(1 - \alpha)$ where α is the required value of the probability of an error of the first kind and β is the required upper limit for $\beta(\theta)$.

Although we have so far assumed that X is a single random variable, all the results remain obviously valid when X is a random vector, i.e., X represents a set of p $(p > 1)$ random variables $X_1, \cdots, X_p$. The only change in the formulas is that the αth observation x_α will have to be replaced by a set $(x_{1\alpha}, \cdots, x_{p\alpha})$ of p values where $x_{i\alpha}$ represents the αth observation on X_i.

A.8.2 Application to Testing the Means of Independently and Normally Distributed Random Variables with Known Variances

Let $X_1, \cdots, X_k$ be k normally and independently distributed random variables with a common known variance σ^2. The mean values $\theta_1, \cdots, \theta_k$ are assumed to be unknown. Suppose that it is required to test the hypothesis that $(\theta_1, \cdots, \theta_k) = (\theta_1{}^0, \cdots, \theta_k{}^0)$. Assume that the zone ω_r of preference for rejection is given by

$$+\sqrt{(\theta_1 - \theta_1{}^0)^2 + \cdots + (\theta_k - \theta_k{}^0)^2} \geqq \delta\sigma$$

where δ is some given positive value. Then the boundary S_r of ω_r is a sphere with center $\theta^0 = (\theta_1{}^0, \cdots, \theta_k{}^0)$ and radius $\delta\sigma$. Let $v(\theta)$ be constant over S_r and equal to the reciprocal of the area of S_r. We shall show that for this weight function conditions (1) and (2) of the preceding section are fulfilled. For this purpose, we shall first prove that the ratio (A:216) is a monotonically increasing function of $(\bar{x}_1 - \theta_1{}^0)^2 + \cdots + (\bar{x}_k - \theta_k{}^0)^2$ where $\bar{x}_i$ is the arithmetic mean of the observations on X_i. In fact, in our case the ratio (A:216) reduces to

$$\text{(A:218)} \quad \frac{c \int_{S_r} e^{-\frac{1}{2\sigma^2} \sum_{i=1}^{k} \sum_{\alpha=1}^{n} (x_{i\alpha} - \theta_i)^2} \, dS}{e^{-\frac{1}{2\sigma^2} \Sigma\Sigma(x_{i\alpha} - \theta_i{}^0)^2}} = ce^{-\frac{1}{2}n\delta^2} \int_{S_r} e^{n \sum_i \frac{(\bar{x}_i - \theta_i{}^0)(\theta_i - \theta_i{}^0)}{\sigma^2}} \, dS$$

where c is equal to the reciprocal of the area of S_r. Let r_x denote $\left| \sqrt{\sum_i \frac{(\bar{x}_i - \theta_i{}^0)^2}{\sigma^2}} \right|$ and let $\rho(\theta)$ $(0 \leq \rho \leq \pi)$ denote the angle between the vector $(\bar{x}_1 - \theta_1{}^0, \cdots, \bar{x}_k - \theta_k{}^0)$ and the vector $(\theta_1 - \theta_1{}^0, \cdots, \theta_k - \theta_k{}^0)$. Then (A:218) can be written as

$$\text{(A:219)} \quad ce^{-\frac{1}{2}n\delta^2} \int_{S_r} e^{nr_x \delta \cos [\rho(\theta)]} \, dS$$

Because of the symmetry of the sphere, the value of (A:219) will not be changed if we substitute $\gamma(\theta)$ for $\rho(\theta)$, where $\gamma(\theta)$ $(0 \leqq \gamma \leqq \pi)$ denotes the angle between the vector $\theta - \theta^0$ and an arbitrarily chosen fixed vector u. From this it follows that the value of (A:219) depends only on r_x.

Now we shall show that (A:219) is a strictly increasing function of r_x. For this purpose we merely have to show that

$$(A:220) \qquad I(r_x) = \int_{S_r} e^{n r_x \, \delta \, \cos \, [\gamma(\theta)]} \, dS$$

is a strictly increasing function of r_x. We have

$$(A:221) \qquad \frac{dI(r_x)}{dr_x} = \int_{S_r} n \delta \, \cos \, [\gamma(\theta)] e^{n r_x \, \delta \, \cos \, [\gamma(\theta)]} \, dS$$

Denote by S'_r the subset of S_r in which $0 \leqq \gamma(\theta) \leqq \pi/2$, and by S''_r the subset in which $\pi/2 < \gamma(\theta) \leqq \pi$. Because of the symmetry of the the sphere we have

$$(A:222) \quad \int_{S''_r} n \, \delta \, \cos \, [\gamma(\theta)] e^{n r_x \, \delta \, \cos \, [\gamma(\theta)]} \, dS$$

$$= \int_{S'_r} n \, \delta \, \cos \, [\pi - \gamma(\theta)] e^{n r_x \, \delta \, \cos \, [\pi - \gamma(\theta)]} \, dS$$

$$= - \int_{S'_r} n \, \delta \, \cos \, [\gamma(\theta)] e^{-n r_x \, \delta \, \cos \, [\gamma(\theta)]} \, dS$$

Hence

$$(A:223) \quad \frac{dI(r_x)}{dr_x} = n \, \delta \int_{S'_r} \cos \, [\gamma(\theta)] (e^{n \, \delta r_x \, \cos \, [\gamma(\theta)]} - e^{-n \, \delta r_x \, \cos \, [\gamma(\theta)]}) \, dS$$

The right-hand side of (A:223) is positive. Hence, we have proved that expression (A:219) or (A:218) is a strictly increasing function of r_x.

We shall now show that $\beta(\theta)$ is constant over any sphere $S_r(d)$ with center θ^0 and radius d and that it decreases monotonically with increasing d. For this purpose let $y_1, \cdots, y_k$ be an orthogonal linear transformation of $x_1 - \theta_1^0, \cdots, x_k - \theta_k^0$ so that $E(y_1) = \sqrt{(\theta_1 - \theta_1^0)^2 + \cdots + (\theta_k - \theta_k^0)^2}$ and $E(y_i) = 0$ $(i = 2, \cdots, k)$. Since $\bar{y}_1^2 + \cdots + \bar{y}_k^2 = (\bar{x}_1 - \theta_1^0)^2 + \cdots + (\bar{x}_k - \theta_k^0)^2$ and since (A:219) depends only on $(\bar{x}_1 - \theta_1^0)^2 + \cdots + (\bar{x}_k - \theta_k^0)^2$, it is seen that the sequence of expressions (A:219) formed for the sequence of integers $n = 1, 2, \cdots$, etc., has a joint distribution which depends only on

$\sqrt{(\theta_1 - \theta_1{}^0)^2 + \cdots + (\theta_k - \theta_k{}^0)^2}$. Hence $\beta(\theta)$ is constant on any sphere $S_r(d)$. Since (A:219) is a strictly monotonic function of r_x, it can be shown that $\beta(\theta)$ is monotonically decreasing with increasing d. Hence, conditions (1) and (2) of the preceding section are fulfilled and we can test the hypothesis that $\theta = \theta^0$ by the sequential probability ratio test based on the ratio (A:218).

If $k = 1$, i.e., if we test the mean value of a single random variable X, the sphere S_r is a null-dimensional sphere consisting of the two points $\theta_1 = \delta\sigma$ and $\theta_2 = -\delta\sigma$ and (A:216) reduces to the ratio of p_{1n} to p_{0n} given by (4:8) and (4:9), respectively, in Section 4.1.4.

A.9 DETERMINATION OF OPTIMUM WEIGHT FUNCTIONS $w_a(\theta)$ AND $w_r(\theta)$ IN SOME SPECIAL CASES OF TESTING COMPOSITE HYPOTHESES

A.9.1 A Class of Cases for Which Optimum Weight Functions $w_a(\theta)$ and $w_r(\theta)$ Can Be Determined by a Simple Procedure

Let $f(x, \theta_1, \cdots, \theta_k)$ denote the distribution of x involving k unknown parameters $\theta_1, \cdots, \theta_k$. Suppose we wish to test the composite hypothesis H_ω that the parameter point θ lies in the subset ω of the parameter space. Let ω_a denote the zone of preference for acceptance and ω_r the zone of preference for rejection. Assume that the boundary of ω_r is a surface S_r. Suppose that it is possible to find two weight functions $v_a(\theta)$ and $v_r(\theta)$ such that

$$\int_{\omega_a} v_a(\theta)\, d\theta = 1, \quad \int_{S_r} v_r(\theta)\, dS_r = 1$$

and that the sequential probability ratio test based on the ratio

(A:224) $\qquad \dfrac{p_{1n}}{p_{0n}} = \dfrac{\displaystyle\int_{S_r} v_r(\theta) \prod_{\alpha=1}^{n} f(x_\alpha, \theta_1, \cdots, \theta_k)\, dS_r}{\displaystyle\int_{\omega_a} v_a(\theta) \prod_{\alpha=1}^{n} f(x_\alpha, \theta_1, \cdots, \theta_k)\, d\theta}$

satisfies the following conditions (for any values A and B): (1) $\alpha(\theta)$ is constant in ω_a; (2) $\beta(\theta)$ is constant over S_r; (3) for any point θ in the interior of ω_r, the value of $\beta(\theta)$ does not exceed the constant value of $\beta(\theta)$ on S_r.

We shall now show that $v_a(\theta)$ and $v_r(\theta)$ may be regarded as optimum weight functions in the sense defined in Section 4.2.2. For this purpose, let $w_a(\theta)$ and $w_r(\theta)$ be any other weight functions and let $\alpha^*(\theta)$

and $\beta^*(\theta)$ be the resulting probabilities of errors of the first and second kinds when $w_a(\theta)$ and $w_r(\theta)$ are used. Since, as has been shown,

(A:225)
$$\int_{\omega_a} \alpha^*(\theta) w_a(\theta) \, d\theta = \frac{1 - B}{A - B}$$

and

$$\int_{\omega_r} \beta^*(\theta) w_r(\theta) \, d\theta = \frac{B(A - 1)}{A - B}$$

hold with good approximation, we see that in ω_a the maximum of $\alpha^*(\theta) \geqq (1 - B)/(A - B)$, and in ω_r the maximum of $\beta^*(\theta) \geqq B(A - 1)/(A - B)$ with good approximation. But if $v_a(\theta)$ and $v_r(\theta)$ are used, it follows from conditions (1), (2), and (3) that (with good approximation) the maximum of $\alpha(\theta)$ in ω_a is equal to $(1 - B)/(A - B)$ and the maximum of $\beta(\theta)$ in ω_r is equal to $B(A - 1)/(A - B)$. Hence these weight functions are optimum in the sense defined in Section 4.2.2.

In some special but important statistical problems one can easily find weight functions $v_a(\theta)$ and $v_r(\theta)$ which satisfy conditions (1), (2), and (3). It will be seen in the next section that such weight functions can easily be constructed when the mean of a normal distribution with unknown variance is being tested. Again, for practical purposes we may let $A = (1 - \beta)/\alpha$ and $B = \beta/(1 - \alpha)$, where α is the required upper bound of $\alpha(\theta)$ in ω_a and β is the required upper bound of $\beta(\theta)$ in ω_r.

A.9.2 Application to Testing the Mean of a Normal Distribution with Unknown Variance (Sequential t-Test)

Let X be a normally distributed random variable with unknown mean θ and unknown variance σ^2. Suppose we wish to test the hypothesis that $\theta = \theta_0$. Furthermore, assume that ω_r is given by the set of all points (θ, σ) for which $\left| \dfrac{\theta - \theta_0}{\sigma} \right| \geqq \delta$, while ω_a consists of all points (θ_0, σ). Then the boundary S_r of ω_r consists of all points (θ, σ) for which $\left| \dfrac{\theta - \theta_0}{\sigma} \right| = \delta$, i.e., it contains the points (θ, σ) for which either $\theta = \theta_0 + \delta\sigma$ or $\theta = \theta_0 - \delta\sigma$.

For any positive value c we define the weight functions $v_{ac}(\sigma)$ and $v_{rc}(\sigma)$ as follows: $v_{ac}(\sigma) = 1/c$ if $0 \leqq \sigma \leqq c$ and equals 0 for all other values of σ. The weight function $v_{rc}(\sigma)$ is equal to $1/2c$ if $0 \leqq \sigma \leqq c$ and $\theta = \theta_0 \pm \delta\sigma$ and equal to 0 otherwise. Let

(A:226)

$$p_{1n} = \int_{S_r} v_{rc}(\sigma) \frac{1}{(2\pi)^{\frac{n}{2}}\sigma^n} e^{-\frac{1}{2\sigma^2}\Sigma(x_\alpha-\theta)^2} \, d\sigma$$

$$= \frac{1}{(2\pi)^{\frac{n}{2}}} \frac{1}{2c} \left(\int_0^c \frac{1}{\sigma^n} e^{-\frac{1}{2\sigma^2}\Sigma(x_\alpha-\theta_0-\delta\sigma)^2} + \frac{1}{\sigma^n} e^{-\frac{1}{2\sigma^2}\Sigma(x_\alpha-\theta_0+\delta\sigma)^2} \right) d\sigma$$

and

(A:227)
$$p_{0n} = \frac{1}{(2\pi)^{\frac{n}{2}}} \frac{1}{c} \int_0^c \frac{1}{\sigma^n} e^{-\frac{1}{2\sigma^2}\Sigma(x_\alpha-\theta_0)^2} \, d\sigma$$

Then

(A:228)
$$\frac{p_{1n}}{p_{0n}} = \frac{\dfrac{1}{2} \displaystyle\int_0^c \dfrac{1}{\sigma^n} \left(e^{-\frac{1}{2\sigma^2}\Sigma(x_\alpha-\theta_0-\delta\sigma)^2} + e^{-\frac{1}{2\sigma^2}\Sigma(x_\alpha-\theta_0+\delta\sigma)^2} \right) d\sigma}{\displaystyle\int_0^c \dfrac{1}{\sigma^n} e^{-\frac{1}{2\sigma^2}\Sigma(x_\alpha-\theta_0)^2} \, d\sigma}$$

We consider the limiting case when $c \to \infty$. Thus

(A:229)
$$\frac{p_{1n}}{p_{0n}} = \frac{\dfrac{1}{2} \displaystyle\int_0^\infty \dfrac{1}{\sigma^n} \left(e^{-\frac{1}{2\sigma^2}\Sigma(x_\alpha-\theta_0-\delta\sigma)^2} + e^{-\frac{1}{2\sigma^2}\Sigma(x_\alpha-\theta_0+\delta\sigma)^2} \right) d\sigma}{\displaystyle\int_0^\infty \dfrac{1}{\sigma^n} e^{-\frac{1}{2\sigma^2}\Sigma(x_\alpha-\theta_0)^2} \, d\sigma}$$

The sequential probability ratio test based on the ratio (A:229) provides a solution to our problem if it can be shown to have the following three properties: (1) $\alpha(\theta, \sigma)$ is constant in ω_a; (2) $\beta(\theta, \sigma)$ is a function of $\left| \dfrac{\theta - \theta_0}{\sigma} \right|$ alone; (3) $\beta(\theta, \sigma)$ is monotonically decreasing with increasing $\left| \dfrac{\theta - \theta_0}{\sigma} \right|$.

To prove these three properties, let $\bar{x}$ denote $\dfrac{\displaystyle\sum_{\alpha=1}^n x_\alpha}{n}$ and S^2 denote $\Sigma(x_\alpha - \bar{x})^2$. Since the joint distribution of a sequence of expressions $\left| \dfrac{\bar{x} - \theta_0}{S} \right|$ corresponding to consecutive values of n depends only on $\left| \dfrac{\theta - \theta_0}{\sigma} \right|$, the first two properties are proved if we show that the ratio (A:229) is a single-valued function of $\left| \dfrac{\bar{x} - \theta_0}{S} \right|$.

First we show that the numerator of the ratio (A:229) is a homogeneous function of $(x_1 - \theta_0,\ x_2 - \theta_0,\ \cdots,\ x_n - \theta_0)$ of degree $-(n-1)$. In fact, making the transformation $\sigma = \lambda t$ we obtain

$$\int_0^\infty \frac{1}{\sigma^n} \left(e^{-\frac{1}{2\sigma^2}\Sigma(\lambda x_\alpha - \lambda\theta_0 - \delta\sigma)^2} + e^{-\frac{1}{2\sigma^2}\Sigma(\lambda x_\alpha - \lambda\theta_0 + \delta\sigma)^2} \right) d\sigma$$

$$= \int_0^\infty \frac{1}{(\lambda t)^n} \left(e^{-\frac{1}{2t^2}\Sigma(x_\alpha - \theta_0 - \delta t)^2} + e^{-\frac{1}{2t^2}\Sigma(x_\alpha - \theta_0 + \delta t)^2} \right) d\lambda t$$

$$= \frac{1}{\lambda^{n-1}} \int_0^\infty \frac{1}{t^n} \left(e^{-\frac{1}{2t^2}\Sigma(x_\alpha - \theta_0 - \delta t)^2} + e^{-\frac{1}{2t^2}\Sigma(x_\alpha - \theta_0 + \delta t)^2} \right) dt$$

This proves that the numerator of (A:229) is a homogeneous function of $x_1 - \theta_0,\ \cdots,\ x_n - \theta_0$ of degree $-(n-1)$. Similarly, it can be shown that the denominator of (A:229) is also a homogeneous function of degree $-(n-1)$. Thus, the ratio (A:229) is a homogeneous function of zero degree in the variables $x_1 - \theta_0,\ \cdots,\ x_n - \theta_0$.

It can be verified that (A:229) is a function of only the two expressions $\Sigma(x_\alpha - \theta_0)^2$ and $\Sigma(x_\alpha - \theta_0)$, i.e.,

(A:230) $$\frac{p_{1n}}{p_{0n}} = \phi[\Sigma(x_\alpha - \theta_0)^2,\ \Sigma(x_\alpha - \theta_0)]$$

Let $v = \left| \sqrt{\Sigma(x_\alpha - \theta_0)^2} \right|$. Since (A:230) is a homogeneous function of zero degree in $x_1 - \theta_0,\ \cdots,\ x_n - \theta_0$, its value is not changed by substituting $(x_\alpha - \theta_0)/v$ for $x_\alpha - \theta_0$. Hence

(A:231) $$\frac{p_{1n}}{p_{0n}} = \phi\left[\sum_\alpha \left(\frac{x_\alpha - \theta_0}{v}\right)^2,\ \frac{\Sigma(x_\alpha - \theta_0)}{v} \right] = \phi\left[1,\ \frac{n(\bar{x} - \theta_0)}{v} \right]$$

Since $\phi[\Sigma(x_\alpha - \theta_0)^2,\ -\Sigma(x_\alpha - \theta_0)] = \phi[\Sigma(x_\alpha - \theta_0)^2,\ \Sigma(x_\alpha - \theta_0)]$, we see that

$$\frac{p_{1n}}{p_{0n}} = \psi\left[\frac{(\bar{x} - \theta_0)^2}{v^2} \right]$$

Since $\dfrac{(\bar{x} - \theta_0)^2}{v^2}$ is a single-valued function of $\left| \dfrac{\bar{x} - \theta_0}{S} \right|$, we have proved

that $\dfrac{p_{1n}}{p_{0n}}$ is a single-valued function of $\left| \dfrac{\bar{x} - \theta_0}{S} \right|$. Hence properties (1) and (2) are proved.

In order to prove property (3) of the sequential probability ratio test based on the ratio (A:229), it is sufficient to show that (A:229) is a strictly increasing function of $\left| \dfrac{\bar{x} - \theta_0}{S} \right|$. Since $\left| \dfrac{\bar{x} - \theta_0}{S} \right|$ is a

strictly increasing function of $\left(\dfrac{\bar{x} - \theta_0}{v} \right)^2$, we have only to show that

(A:229) is a strictly increasing function of $\left(\dfrac{\bar{x} - \theta_0}{v} \right)^2$. The latter

statement is proved if we show that (A:229) increases with increasing value of $|\bar{x} - \theta_0|$ while v is kept fixed. For a fixed value of v the denominator of (A:229) is constant. Thus, we merely have to show that the numerator of (A:229) increases with increasing $|\bar{x} - \theta_0|$ while v is kept fixed. This follows easily from the fact that

$$ e^{\frac{(\bar{x}-\theta_0)\delta}{\sigma}} + e^{-\frac{(\bar{x}-\theta_0)\delta}{\sigma}} $$

is a strictly increasing function of $|\bar{x} - \theta_0|$.

INDEX